Computer Automation
in Manufacturing

Computer Automation in Manufacturing

An introduction

Thomas O. Boucher

Department of Industrial Engineering
Rutgers University
Piscataway NJ
USA

CHAPMAN & HALL

London · Glasgow · Weinheim · New York · Tokyo · Melbourne · Madras

Published by Chapman & Hall, 2–6 Boundary Row, London SE1 8HN, UK

Chapman & Hall, 2–6 Boundary Row, London SE1 8HN, UK

Blackie Academic & Professional, Wester Cleddens Road, Bishopbriggs, Glasgow G64 2NZ, UK

Chapman & Hall GmbH, Pappelallee 3, 69469 Weinheim, Germany

Chapman & Hall USA, 115 Fifth Avenue, New York, NY 10003, USA

Chapman & Hall Japan, ITP-Japan, Kyowa Building, 3F, 2-2-1 Hirakawacho, Chiyoda-ku, Tokyo 102, Japan

Chapman & Hall Australia, 102 Dodds Street, South Melbourne, Victoria 3205, Australia

Chapman & Hall India, R. Seshadri, 32 Second Main Road, CIT East, Madras 600 035, India

First edition 1996

© 1996 Thomas O. Boucher

Typeset in 10/12pt. Times by Thomson Press (I) Ltd., New Delhi
Printed in Great Britain by St Edmundsbury Press, Bury St Edmunds, Suffolk.

ISBN 0 412 60230 X

Apart from any fair dealing for the purposes of research or private study, or criticism or review, as permitted under the UK Copyright Designs and Patents Act, 1988, this publication may not be reproduced, stored, or transmitted, in any form or by any means, without the prior permission in writing of the publishers, or in the case of reprographic reproduction only in accordance with the terms of the licences issued by the Copyright Licensing Agency in the UK, or in accordance with the terms of licences issued by the appropriate Reproduction Rights Organization outside the UK. Enquiries concerning reproduction outside the terms stated here should be sent to the publishers at the London address printed on this page.

The publisher makes no representation, express or implied, with regard to the accuracy of the information contained in this book and cannot accept any legal responsibility or liability for any errors or omissions that may be made.

A catalogue record for this book is available from the British Library

Library of Congress Catalog Card Number: 95–70017

∞ Printed on permanent acid-free text paper, manufactured in accordance with ANSI/NISO Z39.48-1992 and ANSI/NISO Z39.48-1984 (Permanence of Paper). [Paper = Magnum. 70 gsm]*

To My Mother and Father

Contents

Preface	xi
Acknowledgements	xiii

1 Introduction — 1
  1.1 Developments in manufacturing technology for automation — 1
  1.2 A hierarchical model of factory automation — 7
  1.3 System requirements and automatic control technology — 8
  1.4 About this book — 10
  1.5 Summary — 12
  Further reading — 12

2 The architecture of microprocessor-based systems — 13
  2.1 Introduction — 13
  2.2 The transistor and basic digital logic functions — 16
  2.3 Gates and their I/O functions — 23
  2.4 Basic properties of boolean variables — 27
  2.5 Gates, boolean algebra and the design of logic networks — 29
  2.6 An overview of the architecture and operations of a microprocessor-based system — 31
  2.7 Number systems and low-level computer languages — 47
  2.8 Assembly language programming — 57
  2.9 Communications and automatic data transfer — 60
  2.10 Software strategies for I/O communication — 75
  2.11 Summary — 80
  Exercises — 81
  Further reading — 90

3 Sensors and automatic data acquisition — 92
  3.1 Introduction — 92
  3.2 Discrete event sensors — 93
  3.3 Continuous sensors — 100
  3.4 Interfacing a digital controller with an analog world — 106
  3.5 Transducers — 116
  3.6 Summary — 132

		Exercises	132
		Further reading	136

4 Actuators and the performance of work — 137
    4.1 Introduction — 137
    4.2 Concepts of work, force, torque and power — 137
    4.3 Power transmission and reflected forces — 145
    4.4 Basic principles of a DC machine — 148
    4.5 Operating characteristics of DC motors — 152
    4.6 Speed control of DC motors — 155
    4.7 Stepping motors — 158
    4.8 Relay switches and solenoids — 162
    4.9 Fluid actuators — 164
    4.10 Summary — 167
    Appendix 4A: Quantities and their units — 168
    Exercises — 168
    Further reading — 171

5 Control theory — 172
    5.1 Introduction — 172
    5.2 Components of a control system — 172
    5.3 Mathematical characterization and transfer functions — 174
    5.4 Laplace transforms — 176
    5.5 Using transforms to analyze system response — 179
    5.6 Closed loop speed control — 186
    5.7 Position transform of a DC motor — 194
    5.8 Position control of a DC motor — 196
    5.9 Characteristic responses of a closed loop control system — 197
    5.10 Control strategies — 205
    5.11 Typical classes of control system models — 208
    5.12 Digital control — 213
    5.13 Summary — 229
    Appendix 5A: Partial fraction expansion — 230
    Appendix 5B: Solution of a linear differential equation with imaginary roots — 231
    Exercises — 233
    Further reading — 239

6 Programmable logic controllers — 240
    6.1 Introduction — 240
    6.2 PLC hardware — 241
    6.3 Relay circuits and ladder diagrams — 244
    6.4 Sequential control using relay circuits — 247
    6.5 Sequential control using PLCs — 251

6.6	Scan sequence and program execution	258
6.7	Decoupling and program control	259
6.8	Counters and timers	263
6.9	Sequential function charts and GRAFCET	272
6.10	Advanced PLC programming	286
6.11	Summary	290
	Exercises	290
	Further reading	298

7 Supervisory control of manufacturing systems — 299

7.1	Introduction	299
7.2	Production organization and manufacturing system design	301
7.3	Communication architecture and local area networks	308
7.4	Discrete event systems and supervisory controller software design	327
7.5	Some mathematical properties of ordinary Petri nets	335
7.6	Case study: Petri net representation of a machining cell controller	344
7.7	Software specification for a machining cell controller	353
7.8	Imposing priorities in Petri net models	359
7.9	Summary	368
	Exercises	369
	Further reading	371

Appendix: Conversion tables — 372

Index — 373

Preface

The use of industrial PCs and programmable controllers in manufacturing has been growing rapidly over the last twenty years. With their flexibility of programming for different products and processes, capability for real time data gathering for on-line product inspection and process control, and potential for integrating the shop floor with a factory-wide information system, it is likely that this technology will continue to evolve and be used for the forseeable future.

The primary purpose of writing this book is to provide a text with a broad treatment of computer automation that can be used in the third or fourth year of an undergraduate engineering program. The book covers the technology of microprocessor based systems, emphasizing the similarities between microcomputers, microcontrollers and programmable controllers. It covers the use of this technology in automating manufacturing applications at the individual machine level, the manufacturing production line or cell level, and the shop floor level.

The writing of this book was motivated by the need to cover a range of important subjects in an engineering curriculum that is oriented toward manufacturing. Some subjects are so fundamental to manufacturing automation that all engineers interested in manufacturing should have a basic level of understanding of them. Today, every engineer interested in manufacturing should have a grasp of computer architecture and how computer processors are interfaced to electromechanical processes to gather information and control machines. The theories of continuous and discrete control and their applications using imbedded controllers and programmable logic controllers are important ideas for engineers planning to work on the current generation of manufacturing processes. An understanding of the uses of information and the network technologies that support its transmission throughout the factory will be important to the engineers who design and operate the next generation of manufacturing systems.

Typically, one would expect to cover these subjects in individual specialized courses. However, this is seldom done as a requirement at the undergraduate level. This book helps to overcome two problems associated with the individual courses approach. The first problem is a lack of room in the curriculum. In any engineering discipline, there is so much foundation material to teach that the ability to add new material oriented toward the practice of manufacturing is limited. Students may complete a degree in industrial, manufacturing, or mechanical engineering without having studied computer or controller architecture. On the other hand, some curricula may have an entire course on computer architecture and digital logic, but provide no instruction in control theory or local area networks. The second problem is that, regarding manufacturing, these subjects are part of a continuum of related ideas and should be integrated as they are taught. It is the relationship between these technologies that is important to the practising engineer in manufacturing and that relationship should be emphasized in instruction.

This text was written from lecture notes developed for a senior undergraduate course in computer control in manufacturing. The course was designed to be taken by undergraduate students with a background in physics and calculus at the sophomore level. Although it is helpful to have an introductory course in linear circuit theory, it is not essential if the instructor gives a light treatment of the basic equations for passive circuits or if the physics background of the students has touched on elementary concepts of DC circuits.

Due to the introductory nature of this text, I have focused on covering topics to the level required for an understanding of basic principles. Hence, each chapter is supplemented by a further reading list that provides the reader with references to a more exhaustive treatment of the subject matter of the chapter. I have found that it is possible to cover most of the material in this text in one semester even if the students have no background in computer architecture, a subject which usually consumes a third of the instruction time for the course. Instructors with a student population having a background in computer architecture may wish to supplement the remaining subjects with advanced materials from the reading lists.

I have taught this material as part of a four credit course which includes a one credit laboratory that meets for three hours a week. The lab is an important component of instruction because it provides hands-on experimentation that reinforces the concepts of this book. I have built the laboratory around experiments on the inexpensive and popular Motorola 68HC11 microcontroller and the Allen–Bradley 500 series programmable controller, both of which are referenced in this text. Any instructor interested in obtaining a copy of the laboratory manual can write to me at Rutgers University. A solution manual for end-of-chapter problems is also available by writing to me.

Readers should note that this textbook was written for the purpose of instructing students on the principles of computer control. Throughout this book computer programs are used to illustrate control applications in manufacturing. Because of the variable requirements of specific installations, particularly safety requirements, these programs should not be copied for actual installations without appropriate modifications.

Acknowledgements

I would like to thank Renata Joyner and Suzanne Becker for their assistance in typing several earlier versions of the manuscript. Thanks, also, to my colleague E. A. Elsayed, who encouraged me to develop the course on which the book is based, and to my colleague Mohsen Jafari, who provided me with inputs for Chapter 7. I am grateful to Gulgan Alpan, and Wei Zhou for reviewing portions of manuscript, to Peter Brassington and Ramgopal Venkataraman, who helped prepare several of the text drawings and Dan Diana, who provided the cover artwork. Thanks also to my teaching assistant, Ozerk Gogus, who helped me to refine end-of-chapter problems.

Particular acknowledgement is made to the Allen-Bradley Corporation for permission to use material describing the Data Highway local area networks that appears in Chapter 7 (Copyright 1991 Allen-Bradley Company, Inc. Used with permission. Data Highway Plus, DH+, Data Highway 485, DH-485 are all trademarks of Allen-Bradley Company, Inc.) Thanks also to Dr Seymour Melman for permission to use Figure 1.1.

I thank the professional staff of Chapman & Hall for their help. Thanks especially to my Editor, Mark Hammond, and to Helen Hodge for her meticulous work as Copy Editor.

As have many authors, I found that an important person in the realization of a book is a supportive and understanding spouse. I am grateful to my wife, Unn, not only for putting up with my preoccupations while writing this book, but for her active assistance and encouragement toward the end, when I needed it.

Introduction 1

1.1 DEVELOPMENTS IN MANUFACTURING TECHNOLOGY FOR AUTOMATION

This book is a broad introduction to digital control technology and its application in manufacturing automation. The term 'digital control technology' refers to a microprocessor-based system, such as a computer, microcontroller or programmable controller. The term 'manufacturing automation' refers to self-regulation of machines and processes by their controllers. In a wider usage of the term, authors have sometimes also referred to the automation of decision-making processes of management and scheduling of factory production. Here we use the term in the much narrower sense of the control of electromechanical operations.

The advent of self-regulating systems precedes the appearance of the digital computer. Prior to World War II it was common practice to implement some regulatory control in the process industries, such as chemical and petrochemical. Pneumatic controllers and control valves were often implemented in the regulation of temperatures, pressures, levels and flow rates.

By the 1940s, the theory of servomechanisms and the use of electrical circuits to govern processes was a known technology. It is probably fair to say that by then it would have been possible to automate many factory operations, especially in industries which mass-produced the same items repetitively. Although it would have been possible, it would have been uneconomic. Despite the availability of the technology, the cost and lack of implementation standards restricted wide industry use.

The desire and need to use automatic control has rested on a few considerations. The least of these is labor cost. Automatic and computer-controlled machinery almost always requires a certain amount of worker oversight. It is typical to see a highly automated manufacturing line attended by robots with several workers providing oversight in the event of conveyor jams or other uncontrollable breakdowns. It is also true that sophisticated control technology requires more skilled and highly paid individuals in the factory to program and maintain the technology. Although some gains are made in lowering overall labor cost, this alone cannot suffice to justify automatic control.

An important use of control systems is to relieve operators from the monotonous details associated with many industrial processes and to remove them from environmentally unpleasant situations. This was a contributing factor in the early implementation of automatic control in the chemical and petrochemical industries. Another important consequence of the technology of automatic control is the ability to control precisely the variation in dimension and composition of the finished product. This was a contributing factor in the early development of numerically controlled machine tools, which were considered necessary to hold the tight tolerances required in the manufacture of aircraft components. Today, this is one of the significant motivating forces for computer automation in a wide variety of industries that believe that product quality is one of the most important competitive factors in the marketplace. Finally, the arrival of the computer as a replacement for hard wired control systems has introduced the dimension of 'flexibility' or 'flexible response' to corporate manufacturing strategy. Since the computer is a re-programmable controller, it is possible to redefine the control logic of machines and equipment by downloading new programs as needed. Therefore, rapid changeover among products is possible and can shorten lead times in meeting customer demands.

The advent of computer technology has also led to a reconsideration of how a factory should be designed to take maximum advantage of this technology. Concepts such as hierarchical control, distributed control, computer-integrated manufacturing and flexible manufacturing systems, to be discussed later in the book, are largely an outgrowth of the availability of this technology for supporting the decision processes in a factory.

These developments follow a long history of progress in manufacturing technology. It is well to briefly review the history while introducing some of the subjects to be discussed in this book.

1.1.1 The period of mechanization

The modern age of manufacturing began in England in the 18th century with the first industrial revolution. The harnessing of power, either through the newly invented steam engine or the water wheel, made it more efficient to organize work at points of concentration than in small craft shops. Thus was born the idea of the modern factory.

During the 19th century, the industrial revolution entered the United States with the introduction of powered machinery in mechanical fabrication and the standardization of product design for component interchangeability. These technologies are taken for granted today, but at the time they revolutionized manufacturing. Prior to this time it was common practice to fabricate mechanical products one at a time. Each component of the product would be individually fitted to its adjacent components until the entire product was built. The next unit of product would be fabricated in the same manner. No two products contained components that were alike within some specified tolerance.

DEVELOPMENTS IN MANUFACTURING TECHNOLOGY

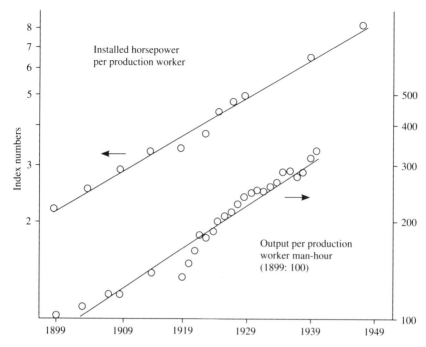

Fig. 1.1 Mechanization and productivity in the manufacturing industries of the US, 1899–1950. (Redrawn from Melman, S., *Dynamic Factors in Industrial Productivity*, published by Basil Blackwell, Oxford, 1956.)

All this changed when it was demonstrated that components of a product could be fabricated identically. Working from mechanical drawings of component specifications, and with relatively precise machine tools, components could now be produced in batches with the assurance that they would fit with mating parts if tolerances were held. It was no longer necessary to manufacture mechanical products on a one-at-a-time basis and the labor required to produce a batch of product could be greatly reduced.

Henry Ford built on these technologies when he applied flow line production to the automobile in 1910. Underlying Ford's innovation was a new mechanism called a 'transfer machine'. A transfer machine was a materials handling device that unloaded and loaded workpieces at each machine in the flow line. The workpiece is loaded and clamped for machining at the first machine and, when machining is completed, unclamped and moved to the next machine by a transfer machine. Though heavily mechanized and inflexible, the Ford factory was a milestone on the road to mass production. When referring to these manufacturing processes, the term 'automation' was coined by the works manager at Ford's factory.

'Detroit automation', as it is known today, emphasized the use of cheap energy, machine innovations, and factory organization to replace muscle power with mechanical horsepower in the performance of work. This form of automation continued through the mid-20th century, resulting in an unprecedented increase in the level of output per factory worker during that period of time.

In a classic study of the causes of mechanization in the United States over the period up through World War II, Seymour Melman of Columbia University documented the impact of mechanization on the rising rate of worker output (Melman, 1956). Some results of his study are shown in Fig. 1.1. Over the period 1900–1949, the level of installed horsepower per production worker grew at a relatively constant rate. The effect of that growth in mechanization was a parallel increase in worker productivity, also shown in Fig. 1.1. His study further shows that the increase in mechanization is caused by the changing relative cost of machinery to workers over that period. In particular, as worker wages rose faster than the cost of machinery, the use of more machinery as an offset to labor was cost-justified. However, instead of observing a reduction in the factory workforce, by 1949 there were over twice as many people employed in US manufacturing than in 1900. This was due in part to the increase in demand for manufactured products, made possible by their low cost from the use of mechanization and mass-production technologies.

1.1.2 Developments in technologies for the self-regulation of processes

In the early 1980s the National Academy of Sciences proclaimed: 'The modern era of electronics has ushered in a second industrial revolution....' (Schmitt and Farwell, 1983). Whereas the first industrial revolution was characterized as replacing muscle power with mechanical power, the second industrial revolution is characterized as replacing brain power with computer power in the decision and control functions in manufacturing.

At the heart of this revolution lies the computer, which could possibly be the foremost technological development in manufacturing of the latter 20th century. The idea of a computer has to be credited to Charles Babbage, an English mathematician. In 1812 he proposed an automatic calculator that could solve difference equations. Although Babbage developed workable designs, his machine was never built.

The modern digital computer takes its origin from a Harvard graduate student, Howard Aikens, who in 1937 worked on the design of a machine that could solve polynomials. His work was later supported by IBM and, in 1944, led to the development of MARK I. MARK I was a 51 ft long electromechanical computer with more than 760 000 parts and 500 miles of wire.

The Harvard computer had many features of a modern computer: an input unit, an output unit, a memory and an arithmetic unit. All instructions had to be programmed and input using a punched tape reader.

DEVELOPMENTS IN MANUFACTURING TECHNOLOGY

The Harvard computer had a competitor. In 1944, the Princeton mathematician John von Neumann collaborated with engineers at the University of Pennsylvania in the development of a computer that worked on the binary number system and that had operating instructions built into the machine itself. Underlying principles for this work appeared in a master's thesis written by Claude Shannon of the Massachusetts Institute of Technology in 1937. This thesis demonstrated the parallel between electrical switching circuits and mathematical logic, which is based on the binary number system.

Supporting the practical implementation of this work was the invention of the transistor at Bell Laboratories in 1948. By replacing the vacuum tube, the transistor made economic the fabrication of binary circuits, called 'logic gates'. By the 1960s, the Texas Instruments Corporation had succeeded in combining several of these gates on a single fabricated chip. Although this appeared to be a relatively dense packing of electronic logic, it was later to become known as 'small scale integration'. By the end of the 1970s it was not uncommon to put 50 000 logic gates on a single chip. This brought hardware costs down dramatically. It has been estimated that a computer having hardware costs of $30 000 in 1960 would have costed about $1000 by 1980.

But automation does not live by hardware alone. Paralleling the advances in hardware were advances in the understanding of how to model manufacturing processes for automatic control. During the 1930s and 1940s, scientists and mathematicians were developing a theory of electromechanical control. The theory of control provides a basis for the design of self-regulating devices. The important concept in the theory is 'feedback'. This is the act of continuously monitoring the output of a process and comparing it to a desired output. The controller alters the input to the process to bring the actual output of the process into conformance with the desired goals. By the 1940s there were several analog devices that performed on this basis.

Since digital computers are discrete devices that monitor a process by taking samples at discrete instances of time, the use of computer technology in the control function needed a control theory based on the sampling of feedback information. This theory began to develop during World War II, when scientists were working on radar systems. A radar system is a naturally sampled system because of the time delay between successive detection of object position. The first important theoretical work on sampled systems began to appear around this time.

The marriage of computer and manufacturing process also began in the late 1940s. The first commercial machines did not employ the concept of feedback. They were programmed using a punched tape that contained instructions for directing the motion of the motors driving the axes of the machine tool. Starting from a known registration point, the bed of a milling machine would be moved through various positions beneath the milling tool in accordance with the instructions on the tape. There was no continuous feedback of positional data, but a proper initial registration of the workpiece, combined with correct instructions, would result in the correct component being manufactured. This process was termed 'numerical control' and it ushered in a great deal of expectation about the potential for automatic control in manufacturing.

The idea of an automatic factory was seriously being entertained in the period just after World War II. In 1946, two radar engineers proposed a prototype automatic factory design in an article that appeared in *Fortune*, a prominent business periodical. With the development of numerical control, expectations increased and, in 1953, 'the automatic factory' became the topic of discussion at The Fortune Round Table, an annual meeting of academic and industrial experts and business executives. The article reporting the discussion at that meeting is interesting to read today in that many of the ideas that are widely discussed today were discussed at that time.

The computer was revolutionary, but applications in manufacturing have tended to be evolutionary. Looking back over the period since 1953, we see a continued application of the computer and control theory to manufacturing and manufacturing processes. However, with a few exceptions, the automatic factory is still not widely realized.

It is not surprising that the implementation of automatic control in manufacturing has moved slowly. For one thing, it requires a higher level of skill on the part of plant engineers and machine operators. Absorbing new technologies into a company while continuing to produce a product is a tenuous affair. There is usually a learning period during which the existing organization learns to apply the new technology. An interesting example of a case in which this fact was recognized by engineering professionals occurred with the introduction of general-purpose programmable controllers into factories. At that time, there were a number of possible languages for programming these controllers. In order to overcome some of this learning lag, early generations of programmable controllers used a programming language that was based on a language already being used by plant electricians for designing hard wired control circuits. This eased the acceptability of this relatively new technology.

Another difficulty with designing self-regulation in a factory is the difficulty of modeling the control problem. The manufacturing environment is very complex and often not well understood. There are many actions of workers along the production line that compensate for unanticipated events during production, e.g. clearing a potential conveyor jam when a product hangs up along a conveyor line. The impact of these informal control actions is not appreciated until one undertakes the automatic control of the line. Since it is difficult to anticipate all occurrences when creating the control logic, software development is often a trial and error affair. This condition is less of a problem when addressing unit operations, i.e. the control of a single machine. However, when multiple machines are being coordinated at some higher level of decision control, this becomes more of an issue. Thus, the challenges and the interesting problems are still there for the next generation of engineers.

1.2 A HIERARCHICAL MODEL OF FACTORY AUTOMATION

In manufacturing plants there are several levels of automation to be considered. Although it is possible to characterize these levels in different ways, we will focus on a four-level classifcation:

1. machine level automation;
2. production line or work cell automation;
3. shop floor automation;
4. plant level automation.

This hierarchy is shown below. The automation problems are different at each level.

Level 4: Plant	Order processing Purchasing Aggregate production planning Accounting
Level 3: Shop floor	Materials management Quality management Shop floor scheduling
Level 2: Work cell* production line	Materials handling Part sequencing Inspection*Statistical process control
Level 1: Machine	CNC machine tools Robots Programmable controllers

The objectives of machine level (level 1) automation are to ensure that the operation of the machine corresponds to the planned sequence of operations, or programmed steps. Typically, the sequence of operations that must be carried out is prescribed and there are no decisions to be made to alter the sequence. Examples of this level of automation are the control of CNC machine tools, industrial robots, and automated assembly machines.

The automation problem at the production line or work cell level (level 2) is normally distinguished from the machine level by the requirement to perform local conditional decision making and coordinate the activities of more than one machine. Typical decisions at this level are controlling the movement of components among machines, coordinating the loading of components on machines in the correct order, and evaluating inspection data on the product as it is being manufactured. The supervising controller at the production line or work cell level is normally not responsible for controlling individual machines through their manufacturing steps. That responsibility is left to the controller at the machine level.

There is a need to coordinate activities among production lines or work cells. This is the automation problem at the shop floor level (level 3). For example, there may be transport devices that bring materials from storage to the work area and return product to storage. Such transport systems are shared among the production lines and work cells of the shop floor. The scheduling and execution of events for these shared resources is a shop floor automation problem since it lies above the control of any single production line or cell.

Automation at the plant level (level 4) is automation in the wider sense of the word. It is primarily concerned with automating overall business decision-making processes or, at a minimum, automating the support of those decision processes through computer information systems. Typical activities in this area include sales and order processing, purchasing, accounting and long-range or aggregate production planning.

The latter chapters of this book focus on the first three levels of automation. We make the distinction between control of unit operations and supervisory control. **Control of unit operations** means the kind of control found at level 1 automation and **supervisory control** means the kind of coordination of activities required at levels 2 and 3.

1.3 SYSTEM REQUIREMENTS AND AUTOMATIC CONTROL TECHNOLOGY

The vague term 'system' is often used to refer to interacting components of a universe, the boundaries of which are defined by the user of the term. Thus, the planetary system refers to the nine known bodies (ignoring asteroids) that circulate about the sun. The planetary system is bounded from the rest of the universe. It most certainly is affected by the rest of the universe but, for the purposes of some analyses, it can be thought of in isolation.

Thus, we refer to a machine tool as a 'system' of electrical and mechanical components. However, it contains motors, which are themselves systems. Furthermore, when multiple machines are brought together with raw materials and a manufacturing plan, another system is created, i.e. a manufacturing system. In deciding on the regulatory control to be applied, it is first necessary to specify the requirements of the system to be regulated.

Without much loss of generality, we can define two specification parameters of systems in the manufacturing arena that are influential in defining the kind of automatic control required. These specification parameters are **time** and **precision**.

SYSTEM AND AUTOMATIC CONTROL TECHNOLOGY

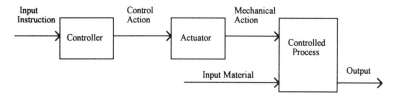

Fig. 1.2 Block diagram of open loop control.

The time specification refers to how quickly it is necessary to update information on the state of the system in order to affect adequate control. We use the term 'states of the system' to refer to the temporal evolution of events that the system undergoes. If the system evolves slowly, which means that its states change slowly, time may not be critical in attending to the control needs of the system. If the system changes state rapidly, a continuous updating of information about the state of the system is required.

The resolution specification refers to the precision with which it is necessary to measure the state of the system. Controlling a system usually requires adjusting the values of one or more of its input variables based on the values of its state variables. If we are controlling the temperature in a room and wish to maintain the temperature to within one degree, some fairly precise resolution of the state of the system is required. If we are controlling the lamp in a room and we are only interested in two conditions, lamp on and lamp off, there is little measurement resolution required in measuring the state of the process. We only need to decide in which of two states the system resides.

Time and resolution are influential in determining the most appropriate automatic controls for a particular system. An **open loop** control system is one in which the state of the system is not monitored at all. Figure 1.2 is a block diagram of a typical open loop controller. In this arrangement, an external input is provided to the controller, which controls an actuator. The actuator, in turn, drives the process. The precision of an open loop system cannot, strictly speaking, be assured. It has usually been used in the sequencing of events in time that requires the use of ON/OFF signals. This kind of control has been used extensively with home appliances, e.g. washing machines. Although it was also used in early generations of numerically controlled machine tools, today's machine tool controllers are closed loop.

Figure 1.3 is a block diagram that illustrates a **closed loop** controller. As the controller executes its instructions, the state of the process is monitored by a sensing device to determine whether or not the system is evolving correctly. When errors are detected between the desired state of the system and the actual state of the system, a programmed procedure is used to bring the actual state of the system into conformance. The closed loop design is thought of as most closely associated with self-regulating devices.

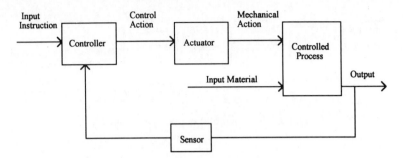

Fig. 1.3 Block diagram of closed loop control.

When time and high resolution are critical to control the system, the closed loop model is appropriate. The term **continuous control** describes the requirement to monitor the system constantly and adjust the input parameters of the system. The control of a servo motor positioning a robot arm and the control of the time and temperature of a chemical reaction are typical applications for continuous control.

When system states change at discrete instances of time, the system is often referred to as a **discrete event system**. Such systems often do not require time-critical control specifications or high resolution measurement. An example of this is the control of the sequence of events for filling material in a hopper. When the hopper is empty, a sensor at the bottom of the hopper registers this state and the controller opens a valve to dump material into the hopper. When the hopper becomes full, a sensor at the top of the hopper registers this state and the controller closes the valve. This is a closed loop system because the controller is receiving feedback from the two sensors that describe the states of the system. However, there are only two states that are of interest: empty and full. The control problem is not concerned with all the intermediate states. The term **discrete control** describes the requirements of this system.

In the latter chapters of this book we will describe problems of both continuous and discrete control. It will be seen that the methods of analysis and specification of control logic differ considerably.

1.4 ABOUT THIS BOOK

This book brings together a study of the technology of the digital controller and some of the manufacturing control problems to which it is applied. It is introductory in nature and meant to be very broad in its coverage. The topics that comprise each of the chapters are themselves the subjects of entire books. Therefore, this book is intended to convey a broad appreciation for many of the issues involved in automation while not exhausting the subject. Each chapter includes a bibliography of advanced material for further study of each topic.

Chapter 2 describes the architecture of microprocessor-based systems. These include personal computers and industrial controllers. The means by which computers process information and interact with the outside environment when communicating and acquiring information is fundamental in understanding their use with sensors and actuators.

In Chapter 3, Sensors and Automatic Data Acquisition, we discuss the various sensors that are used to acquire information in factory automation. Of particular interest is explaining how the digital computer is interfaced to an analog world using analog to digital and digital to analog converters.

Just as Chapter 3 addresses the acquisition of data on the input side, Chapter 4 addresses the use of data to drive processes on the output side. In this chapter we describe actuators that computers control in the performance of work. These include motors, switches, solenoids and fluid valves.

Chapter 5, Control Theory, brings together the preceding topics by describing how these hardware components can be implemented with appropriate software for the purpose of continuous control. Here we introduce methods of process modeling and the concepts of feedback and control strategies. Both analog and digital modeling techniques are described.

Chapter 6, Programmable Logic Controllers, covers special-purpose industrial controllers that are primarily used in discrete control applications. Both the traditional method of ladder logic programming and the newer method of using graphical programming languages are described.

Chapters 2 through 6 focus mainly on the control of unit operations. In Chapter 7, Supervisory Control of Manufacturing Systems, control at the level of interacting unit operations is discussed. Here we describe communication requirements for coordinating the activities of multiple machines and introduce the modeling power of Petri nets as a means of analyzing system behavior and controller logic.

The purpose of this book is to describe foundation material in the application of computers to the automation of manufacturing processes. It is not intended to describe the details of specific manufacturing processes and machines, such as CNC milling machines and robots. Although these devices are used as examples of computer control, their operations and their special-purpose programming languages are not covered in this text. Textbooks on the subject of computer-aided manufacturing usually address these and other topics appropriate to the application of these machines.

This book is also not intended to describe the information processing and managerial control issues of factory automation. Such issues involve factory database design, materials requirements planning, automatic identification, and information control. Textbooks on the subject of computer-integrated manufacturing usually address these and other topics appropriate to the overall information architecture of the manufacturing plant.

1.5 SUMMARY

Manufacturing automation has proceeded incrementally since the early 20th century. The advent of the computer and developments in sensors, digital communication and programming languages have greatly enhanced the possibilities for automation and computer integration in factories. The parallel developments in modeling manufacturing processes and systems have provided the basis for designing control systems and control software. Although, for economic reasons, the 'automatic factory' is still an evolving concept, there is a clear tendency on the part of companies to invest in this technology and to continue to push the frontiers of automatic control. As the development cost of automation software falls, we can anticipate further incremental applications throughout manufacturing. In this book we describe some of the basic technologies such applications will use.

FURTHER READING

Anon. (1953) The automatic factory, *Fortune*, October.
Cannon, D. L. and Luecke, G. L. (1979) *Understanding Microprocessors*, Texas Instruments, Inc., Dallas, Texas.
Leaver, E. W. and Brown, J. J. (1946) Machines without men, *Fortune*, November.
Melman, S. (1956) *Dynamic Factors in Industrial Productivity*, Basil Blackwell, Oxford.
Schmitt, N. M. and Farwell, R. F. (1983) *Understanding Electronic Control of Automation Systems*, Texas Instruments, Inc., Dallas, Texas.
Williams, M. R. (1985) *A History of Computing Technology*, Prentice-Hall, Englewood Chiffs, New Jersey.

The architecture of microprocessor-based systems

2

2.1 INTRODUCTION

The purpose of this chapter is to describe the fundamentals of the operation of a microprocessor-based system, such as a microcomputer or a programmable industrial controller. The word 'fundamentals' should be emphasized, since it is not possible in one chapter to cover the range of topics that should be included for a full understanding of microprocessors and microcontrollers. However, for those readers who have not had a full course on this subject, this chapter will be sufficient to give an understanding of this very important and interesting technology.

First of all, we should define three related terms: microprocessor, microcomputer, and controller (or microprocessor-based controller). A **microprocessor** is a digital integrated circuit that contains the digital functions necessary to 'process' information. It is, in effect, the control center of the digital system of which it is a component; for example, a microcomputer. A microprocessor has a built-in instructions set, which is part of the electronic design of the device. The instruction set tells the device how to control the operation of the system and how logically to execute user-written programs.

Microprocessors come in various sizes and capabilities. One classification is based on the number of bits of data that can be manipulated simultaneously. Typical classes are 8-bit, 16-bit and 32-bit. In this chapter we will focus our attention on 8-bit architecture. In general, the basic concepts are extendible to 16 and 32-bit devices.

By itself, the microprocessor is not very functional. In order for it to perform, it must be used in conjunction with other devices. These other devices allow it to interact with the outside world and give it a place to store user programs, data, and the results of computation. When these devices are configured as shown in Fig. 2.1, the result is a microcomputer.

A **microcomputer** is a small digital computer which uses a microprocessor as its control processing unit and contains all the functions of a computer. Hence, a microcomputer is composed of a microprocessor, memory, input and output devices, and wired connections (called buses) between system components. Typical input devices are keyboards and disk drives; typical output devices are

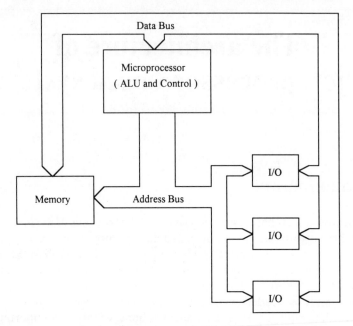

Fig. 2.1 Components of a microcomputer.

monitors and printers. The important feature of the microcomputer is that it can be programmed by a user with a sequence of instructions to be executed.

A **controller**, or microprocessor-based controller, can be subdivided into two categories: programmable industrial controllers and microcontrollers. A **programmable controller**, as defined by the National Electrical Manufacturers' Association (NEMA), is 'a digitally operating electronic apparatus which uses a programmable memory for the internal storage of instructions for implementing specific functions such as logic, sequencing, timing, counting, and arithmetic to control, through digital or analog input/output modules, various types of machines or processes' (Schmitt and Farwell, 1983). The inclusion of programming, storage, logic, and arithmetic capability makes the programmable controller sound very much like a microcomputer. In fact, it is a special form of microcomputer and uses essentially the same internal architecture.

Microcomputer systems were traditionally never designed for use in harsh industrial environments, although they have been successfully used there. The microcomputer, with appropriate input/output (I/O) devices, could function as a programmable controller. However, it would require special construction to be able to withstand the rough handling, range of operating temperatures, and electrical noise typical of the industrial environment. The solution found preferable by manufacturers of robots, CNC machine tools, and other machinery requiring programming was to tailor a microprocessor-based system to the specific needs of its application. This led to the creation of programmable controllers with built-in digital and analog I/Os that could be directly interfaced to industrial processes. As in the case of a microcomputer, they are equipped with memory and programmable inputs and outputs. Typically, the programming language is tailored to the needs of the mechanical process, e.g. for machining or robot control. A more general-purpose programmable controller is called a **programmable logic controller** (PLC). This class of controller is most often programmed in a language known as ladder logic, which is based on the operation of relay circuits. We shall focus on programmable logic controllers in a later chapter.

Unlike a programmable logic controller, a **microcontroller** is a microprocessor-based system which implements the functions of a computer and a controller on a single chip. The device is typically programmed for one application and dedicated to a specific control function, such as the control of the gas/air mixture in a combustion process or the control of temperature in an oven. Microcontrollers are found in automobiles, aircraft, medical electronics and home appliances, as well as other applications. The virtue of the microcontroller is that it is small and can be embedded in an electromechanical system without taking up much room.

In our discussions of programmable microprocessor-based systems, we will be describing the general technology of microcomputers, programmable controllers and microcontrollers. We will often use the terms interchangeably since these machines are based on the same technology. The important distinction to keep in mind is that microcomputers are designed with computational capability in mind whereas programmable controllers and microcontrollers are designed with input/output capability in mind. The important requirements for a microcomputer are memory size and speed of execution. Important requirements for a controller are the number of digital I/O points, the number of input and output analog channels, and the presence of special registers for counting and timing events.

Figure 2.2 shows two controllers. Figure 2.2(a) is a Motorola 68HC11 microcontroller mounted on a prototype board. The functions of the microcontroller are completely designed into the chip. This chip contains very little user programmable memory (approximately 256 bytes of RAM and 512 bytes of electronically programmable EPROM). However, there are a total of five parallel I/O ports, two serial ports, eight analog input channels, plus counters and timers. All on a single chip! The 60 pin wire wrap area is used to make these functions accessible to the outside.

Figure 2.2(b) is an Allen Bradley series 500 programmable logic controller. This machine is typically used to program applications in manufacturing plants. It is rack mounted, with slots into which a user can install modules to configure the controller to a desired application. Here it is shown with modules for digital input, digital output, and analog input/output. The programmable logic controller is the standard controller for flexible

ARCHITECTURE OF MICROPROCESSOR-BASED SYSTEMS

Fig. 2.2 Examples of (a) a microcontroller and (b) a programmable logic controller.

automation in a factory environment. It can be programmed for a particular application and later reconfigured and reprogrammed as control requirements change.

2.2 THE TRANSISTOR AND BASIC DIGITAL LOGIC FUNCTIONS

Components of a microprocessor-based system are packaged in units called integrated circuits (ICs); see Fig. 2.3. These ICs contain individual logic functions in the form of electronic circuits that are accessible on the outside through a pin structure.

The basic element in the IC device family is the transistor. The details of the construction of a transistor are not of particular interest here, but its function-

THE TRANSISTOR AND BASIC DIGITAL LOGIC FUNCTIONS

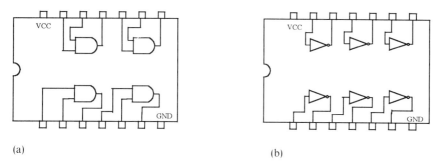

(a)   (b)

Fig. 2.3 Typical integrated circuit (IC) construction. (a) 7408 quad AND gate; (b) 7404 inverter.

ing and its logical characteristics are. In the interest of completeness, we will give a basic introduction to the functioning of a transistor.

The **transistor** has three components: the base, the emitter, and the collector. These components are made of semiconductor material. Semiconductor material, as the name implies, has a higher conductivity than an insulator, such as glass, but a lower conductivity than a conductor, such as copper. Two common semiconductor materials are silicon and germanium. To make these materials useful as semiconductors they are subjected to a process known as 'doping'. The doping process will alter the atomic structure of the material and create either free electrons in the material or free holes (vacancies in the atom where electrons could exist). In the first instance an N-type (negative) material has been created. In the second case a P-type (positive) material has been created.

Transistors are made up of both N-type and P-type material. Figure 2.4 illustrates the two types of bipolar junction transistors. The NPN transistor

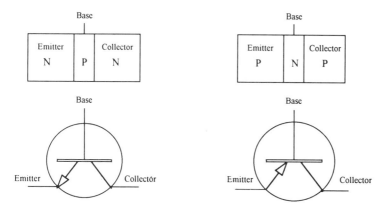

Fig. 2.4 Transistor construction and symbols.

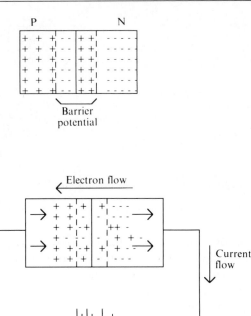

Fig. 2.5 Barrier potential and current flow. (a) Electron and hole potential at junction; (b) current flow when forward biased.

has P-type material sandwiched between two N-types. The PNP transistor is an N-type between two P-types. The accompanying diagrams are the usual schematic representations of these two devices.

The behavior of a transistor is the result of the activity at the interface, or junction between N-type and P-type material. The electrons in the N-type material are attracted to the vacancies, or holes in the P-type material, and vice versa. When the materials are fused, a certain amount of drifting of electrons into the P-type material takes place. This drifting will eventually cease and remaining electrons will be prevented from drifting by a negative potential caused by the electrons that have already occupied holes at the junction. Likewise, the holes created at the junction in the N-type material will set up a positive potential. The effect of these actions is to set up a barrier potential at the junction that prevents further electrical activity. This is illustrated for a P–N junction in Fig. 2.5a.

A P–N junction can be made to conduct by applying a voltage across the material. This is illustrated in Fig. 2.5b. If a voltage is applied with the voltage source as shown, the positive voltage will tend to repel holes in the P material toward the potential barrier. Similarly, the negative voltage will tend to repel

THE TRANSISTOR AND BASIC DIGITAL LOGIC FUNCTIONS

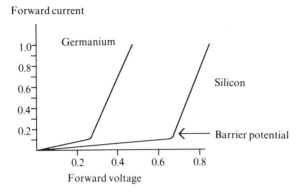

Fig. 2.6 Characteristic curves for forward biased PN junction using germanium and silicon materials.

electrons toward the potential barrier. As the voltage is increased, the potential barrier will be overcome and electrons and holes will begin to flow, setting up a current flow in the indicated direction. The relationship between the applied voltage and the magnitude of the current flow is illustrated in Fig. 2.6 for germanium and silicon semiconductor material. Up to the barrier potential voltage, very little current will flow. Once the barrier potential is broken down, the semiconductor conductivity is dramatically increased. Typical barrier potentials for germanium and silicon are 0.3 volts and 0.7 volts, respectively. The act of applying a voltage to overcome the barrier potential and to cause conduction is called **forward biasing**.

If the polarity of the voltage source is reversed, the junction is said to be **reverse biased**. This is shown in Fig. 2.7. Reverse biasing tends to enlarge the barrier potential. The electrons in the N material will be drawn away from the P–N junction, as will the holes in the P material. This will tend to increase the resistivity at the junction and impede current flow.

The forward and reverse biasing of N and P materials is the basis of transistor operations. Figure 2.8 illustrates the case for an NPN transistor in an important configuration known as a common emitter configuration. With the voltages as shown, the emitter/base junction is forward biased, as in Fig. 2.5, and the collector/base junction is reverse biased, as in Fig. 2.7. Electrons from the emitter will flow into the base region because of the forward biasing. When they arrive in the base area, the potential across the collector/base junction caused by its reverse bias will tend to sweep the electrons into the collector. In fact, only a small number of electrons flow in the base circuit. The majority are drawn to the positive valence of the collector caused by the depletion of electrons during the act of reverse biasing. The magnitude of the electron flow in the collector circuit is in part a function of the voltage applied in the base circuit. As the base voltage is increased, more electrons will tend to

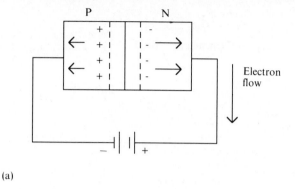

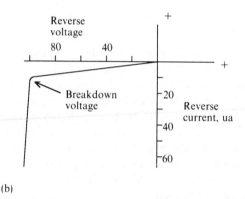

Fig. 2.7 Reverse biasing. (a) Electron flow when reversed biased; (b) typical characteristic curve for a reversed biased PN junction.

flow out of the emitter into the base and they will be swept into the collector in larger numbers.

Figure 2.8b is Fig. 2.8a transformed to a schematic representation of the device. Note the duality between the two diagrams. Voltage V_{BE} is forward biasing the emitter/base junction. The collector/base junction is reverse biased. The flow of current is from collector to emitter since, by convention, current flows in the direction opposite to that of the electrons.

The effect of this behavior of the transistor can be seen in the family of operating characteristic curves. A typical family of these curves is shown in Fig. 2.9. These curves are generated by applying a voltage across the base circuit, causing the base current, I_B, to flow. With the collector circuit voltage fixed, increasing the base current will lead to more free electrons entering the base and being swept away into the collector. This, in turn, increases the magnitude of the collector circuit current, I_C. These curves illustrate the

THE TRANSISTOR AND BASIC DIGITAL LOGIC FUNCTIONS

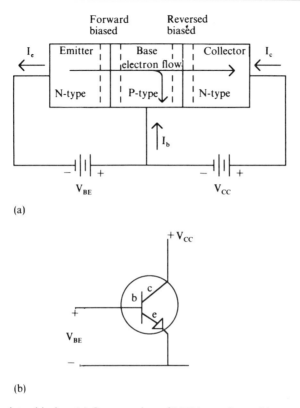

Fig. 2.8 Transistor biasing. (a) Construction of NPN transistor; (b) common emitter configuration.

amplifying characteristic of the transistor. For a small increase in the base current, there occurs a larger increase in the collector current. The ratio of the increase in collector current to the increase in base current is called the **gain**:

$$\beta = \frac{\Delta I_C}{\Delta I_B}$$

where β = the gain of the transistor. Transistors are current amplifying devices.

When a transistor is used in a digital circuit, it is employed as an on/off switch. Figure 2.10 illustrates the method of operation. If V_{input} is slightly negative, applying a reverse bias to the emitter/base junction, a barrier potential will be established and electrons will not flow from emitter to base. The NPN transistor is off. When the base is forward biased and the barrier potential between base and emitter is overcome, the transistor is on, or conducting.

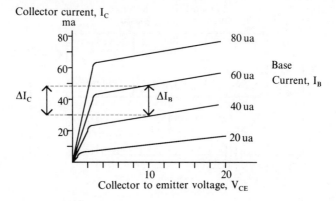

Fig. 2.9 Common emitter characteristic curves.

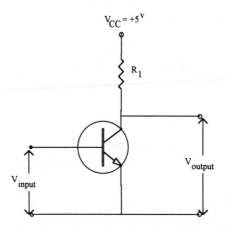

Fig. 2.10 Switching behavior of a transistor circuit.

A standard circuit that illustrates the usefulness of this switching behavior is illustrated in Fig. 2.10. A 5-volt source and a resistor are placed in series with the transistor. When reverse bias is placed on the base of the transistor, the transistor does not conduct and its resistance appears to be very large in comparison to R_1. Therefore, if the voltage drop were measured between source and ground, it would appear almost entirely across the transistor. This is so because the voltage drops across the transistor according to the formula:

$$V_O = \frac{R_T}{R_T + R_1} V_I$$

where R_T is the effective resistance of the transistor.

GATES AND THEIR I/O FUNCTIONS

When the base is forward biased and the transistor is conducting, its resistance drops to a very low value relative to R_1. Consequently, the voltage drop across R_1 is close to 5 volts and the voltage drop across the transistor falls close to 0 (actually, about +0.4 volts).

If you take the point of view that V_O is the output of interest of this circuit, and arbitrarily designate an output above 2.5 volts as a logical level 1 and an output below 2.5 volts as a logic level 0, you have a simple two-state logic device whose output is conditioned on its input. If we refer to the two states on the input arbitrarily as 0 (no forward bias) and 1 (forward bias), a simple input/output relationship can be expressed as shown below (where V_I = input; V_O = output). Such tables are referred to as **truth tables** and are useful tools for expressing the input/output function of a logic circuit.

V_I	V_O
0	1
1	0

When transistors are combined in integrated circuits with other electronic devices, they can form digital circuits. The simplest of the digital circuits is called the **logic gate**. The logic gate performs logical manipulation of one or more logical variables. There are basically three gates: the AND gate, the OR gate, and the inverter, or NOT gate. When the inverter is combined with the AND gate, it yields a NAND gate; when it is combined with the OR gate, it yields the NOR gate. We will describe these gates as a necessary foundation for the discussion of digital circuit operation.

2.3 GATES AND THEIR I/O FUNCTIONS

An **AND gate** is a digital device that performs the logical product or conjunction of two or more logical variables. For the logical AND of two variables we use the notation $A \cdot B$.

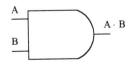

Fig. 2.11 IEEE symbol for AND gate.

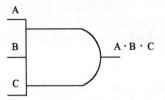

Fig. 2.12 AND three variables.

Figure 2.11 shows the standard IEEE representation of an AND gate and its truth table is represented below ($X = A \cdot B$). From this you can see that the

A	B	X
0	0	0
0	1	0
1	0	0
1	1	1

AND operation requires that both A and B are true (logical 1) in order for X to be true (logical 1). All other combinations are 0. You can AND any number of variables; for example, Fig. 2.12 and its corresponding truth table show the logical AND for three variables ($X = A \cdot B \cdot C$).

A	B	C	X
0	0	0	0
0	0	1	0
0	1	0	0
0	1	1	0
1	0	0	0
1	0	1	0
1	1	0	0
1	1	1	1

An **OR gate** is the logical sum or disjunction of two or more logical variables. For the logical OR of two variables, A or B, we use the notation: $A + B$.

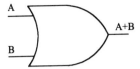

Fig. 2.13 IEEE symbol for OR gate.

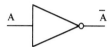

Fig. 2.14 IEEE symbol for NOT gate.

Figure 2.13 is the symbol for an OR gate and its truth table is shown below ($Y = A + B$). From the table it can be seen that the interpretation of logical OR is that if either A or B or both are true, Y is true.

A	B	Y
0	0	0
0	1	1
1	0	1
1	1	1

The **NOT gate**, or inverter, produces the logical complement of a variable. We use the notation $\bar{A}$ for NOT A. The IEEE symbol is shown in Fig. 2.14 and its truth table is shown below.

A	$\bar{A}$
0	1
1	0

These three gates, the AND, OR, and NOT, can be combined to form other logical devices. For example, the **NAND gate** is an AND gate followed by an inverter. It gives the logical complement of a logical AND operation. The transition is illustrated by Fig. 2.15a. For a NAND gate we use the notation:

$$X = \overline{A \cdot B}$$

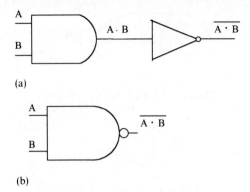

(a)

(b)

Fig. 2.15 IEEE symbol for NAND gate. (a) AND gate followed by inverter; (b) NAND gate.

The NAND gate is important enough to have its own symbol, shown in Fig. 2.15b. The truth table for a NAND gate is given below.

A	B	X
0	0	1
0	1	1
1	0	1
1	1	0

When an OR gate is followed by an inverter, you get the logical complement of a logical OR operation. This is called a **NOR gate** and its notation is:

$$X = \overline{A + B}$$

The symbol for a NOR gate is given by Fig. 2.16 and its truth table appears below ($Y = \overline{A + B}$).

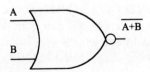

Fig. 2.16 IEEE symbol for NOR gate.

A	B	Y
0	0	1
0	1	0
1	0	0
1	1	0

2.4 BASIC PROPERTIES OF BOOLEAN VARIABLES

In order to understand the functioning of some of the more important components of a microprocessor-based system, it is useful to show what happens when different logical devices are combined. Combinations of logic devices follow the elementary properties of boolean variables. Hence, we will first discuss the important properties of a boolean variable.

Several important properties of any boolean variable follow the AND, OR, and NOT operations. If A is a boolean (0, 1) variable, then the following basic laws are true:

$$A \cdot 1 = A$$
$$A \cdot 0 = 0$$
$$A \cdot A = A$$
$$A \cdot \bar{A} = 0$$
$$\bar{\bar{A}} = A$$

$$A + 0 = A$$
$$A + 1 = 1$$
$$A + A = A$$
$$A + \bar{A} = 1$$

The validity of the above laws can be tested by allowing A to take on any boolean value (0, 1). For example, in $A \cdot 1 = A$, the relationship holds whether $A = 0$ or $A = 1$.

There are four important combinational properties of boolean variables. They are the commutative, associative, distributive and absorptive properties.

Let A, B and C be boolean variables. The **commutative property** states that the order of operation of two boolean variables does not matter. Symbolically,

$$A \cdot B = B \cdot A$$
$$A + B = B + A$$

The **associative property** states that, if there are more than two boolean variables and their logical operations are the same, the order of operation does not matter. This is shown symbolically as follows:

$$A \cdot (B \cdot C) = (A \cdot B) \cdot C = A \cdot B \cdot C$$
$$A + (B + C) = (A + B) + C = A + B + C$$

When three boolean variables are subject to different logical operations, the associative property does not hold.

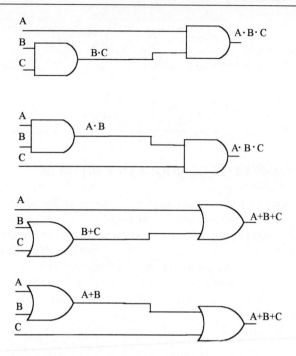

Fig. 2.17 Equivalent circuits by the associative property.

For a boolean property, there is an equivalent digital circuit representation. For example, using IEEE symbolic representation, the equivalent circuits illustrating the associative properties as given above are shown in Fig. 2.17.

The **distributive property** illustrates equivalent ways of combining a boolean variable with two other boolean variables combined using the dual operators. Two symbolic statements of the distributive property are:

$$A \cdot (B + C) = (A \cdot B) + (A \cdot C)$$
$$A + (B \cdot C) = (A + B) \cdot (A + C)$$

The distributive property is a useful vehicle to illustrate how boolean algebra is used in the design of digital circuits. Figure 2.18 shows two circuits that are functionally equivalent according to the distributive property. However, the design for circuit 2 requires an additional OR gate and, therefore, is less design-efficient (cost-effective) than circuit 1. This illustrates how a designer can use boolean algebra to rationalize the design.

The **absorptive property** states that if you perform a logical operation on a variable that has just been combined with another variable using the dual

GATES, BOOLEAN ALGEBRA AND DESIGN OF LOGIC NETWORKS

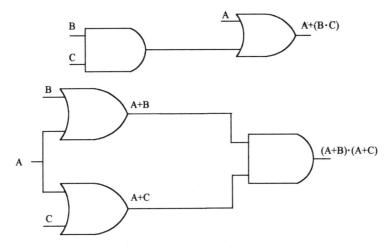

Fig. 2.18 Two equivalent circuits by the distributive property.

operation, you get the variable itself as a resultant. Symbolically:

$$A \cdot (A + B) = A$$
$$A + (A \cdot B) = A$$

What are the economic implications of this for circuit design?

Finally, another relationship that has proven useful in circuit design is known as De Morgan's theorem. De Morgan's theorem proves the equivalence of the following boolean operations:

$$A \cdot B \cdot C = \overline{\overline{A \cdot B \cdot C}} = \overline{\overline{A} + \overline{B} + \overline{C}}$$

2.5 GATES, BOOLEAN ALGEBRA AND THE DESIGN OF LOGIC NETWORKS

The output of a combinational network is a function of its inputs. When networks are designed the desired input/output relationship is stated first, followed by the application of design rules to construct the gate structure for the device. A procedure that works well in the design of combinatorial networks is given by the following three steps:

- Step 1: Construct a truth table for the desired input/output relationship.
- Step 2: For each true statement in an output, form the logical product (AND) of the inputs.
- Step 3: For each output having true statements, form the logical sum (OR) of the logical products of Step 2.

The result of this procedure is the boolean statement of an equivalent digital circuit to perform the desired input/output functions. It is useful to illustrate this procedure by example.

EXAMPLE 2.1

There is a useful digital circuit that performs what is called an **exclusive OR** operation. In an exclusive OR operation, the output of the circuit is false only when the inputs are at the same logic level. Otherwise the output is true. Develop the combinatorial network of an exclusive OR circuit.

Answer

Step 1: The truth table for an exclusive OR (EOR) operation is as follows: $X = A(EOR)B$

	Input		Output
	A	B	X
	0	0	0
	0	1	1
	1	0	1
	1	1	0

Step 2: There are two outputs that are true. We will form their logical products using the following convention:

$$A = 1 \quad B = 1$$
$$\bar{A} = 0 \quad \bar{B} = 0$$

Then, the logical ANDs for $X = 1$ are:

$$\bar{A} \cdot B \quad A \cdot \bar{B}$$

Step 3: The logical sum (OR) of the logical products of Step 2 relates the inputs to the desired output. Hence

$$X = (\bar{A} \cdot B) + (A \cdot \bar{B})$$

This is the boolean statement for an exclusive OR circuit and we can construct its combinational network from the boolean statement. This is shown in Fig. 2.19a.

The exclusive OR circuit performs an important operation in digital circuits and it has been given its own gate symbol. Figure 2.19b shows the symbolic representation of an exclusive OR gate.

ARCHITECTURE AND OPERATIONS OF MICROPROCESSOR SYSTEM 31

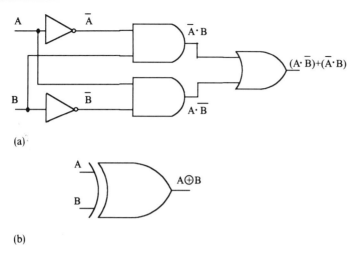

Fig. 2.19 The exclusive OR gate. (a) Exclusive OR circuit; (b) IEEE symbol.

This brief introduction to digital circuit design is a sufficient basis to understand the functions of electronic gates. In what follows we will discuss the functions of the microprocessor-based system itself, introducing the concepts we have just learned as needed to describe some specific operations.

2.6 AN OVERVIEW OF THE ARCHITECTURE AND OPERATIONS OF A MICROPROCESSOR-BASED SYSTEM

A discussion of the architecture and operation of a microcomputer is a good vehicle for introducing the general class of programmable microprocessor-based systems, including programmable controllers. Although there are differences based on manufacturer's specifications, we can describe the fundamentals of a typical microcomputer using the block diagram of Fig. 2.20. We shall first discuss each of the components of the system and their interrelationships. The major components of the system are the memory, the central processing unit (CPU), which is the microprocessor, and the I/O ports. These are linked by a bus structure, which carries the communications signals.

The CPU is the brain of the system; it both controls the system operation and executes the programmed instructions. The memory is a storage device for program instructions. The I/O ports are hardware devices that facilitate the passage of information in and out of the microcomputer. The bus structure, which ties components together, comprises transmission lines over which communication takes place. We shall begin a description of Fig. 2.20 by elaborating on the block labeled 'memory' at the lower left-hand corner of the figure.

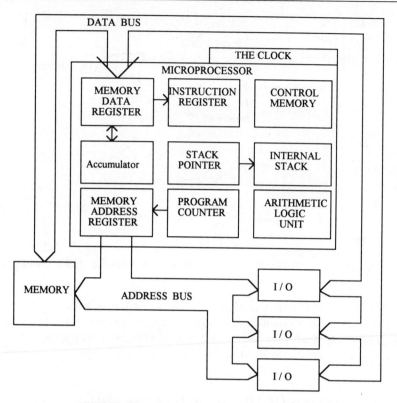

Fig. 2.20 Key components of a microcomputer.

Computer memory is of two generic types: RAM and ROM. RAM, or **random access memory**, is volatile memory. The microprocessor can write information to RAM and read information from RAM. RAM is composed of sequential combinatorial devices that will temporarily retain data as long as power is supplied to the RAM chips. We shall discuss the nature of these devices in section 2.6.2.

ROM, or **read only memory**, is a permanent storage device from which information can only be read. ROM does not support information or program storage during the operation of the machine. It is used for permanent instructions, such as the operating system of the computer. ROM is non-volatile, i.e. the information contained in ROM is not lost when power is removed from the chip.

ROM is programmed at the factory and cannot be reprogrammed. However, users of computer systems such as programmable controllers often need to reprogram an application in the field and provide a permanent storage of the program in memory. The EPROM was developed for this purpose. The

ARCHITECTURE AND OPERATIONS OF MICROPROCESSOR SYSTEM

EPROM, or **erasable programmable read only memory**, is an electronically alterable chip that can be erased and reprogrammed. Therefore, it is a semi-permanent information storage device that can be programmed by a user in this field.

It is convenient to think of memory (ROM or RAM) as consisting of a series of addresses in which information is stored. Figure 2.21a is a characterization of a sequence of memory addresses, each containing four bits of information. A bit (binary digit) is the basic unit of information; each bit provides enough information to distinguish between two states (1, 0; true, false; on, off).

The address bus and data bus work in conjunction with memory for the storage and retrieval of information. The address bus is a group of parallel connections (wires) between memory and the microprocessor. A specific address is accessed by the microprocessor by placing the binary number of that address on the address bus. When a particular address is accessed, the data buffer can read the contents of that address or write information into that address (RAM only). This information passes between memory and the microprocessor over the data bus.

A more detailed description is provided in Fig. 2.21b. This shows a particular kind of ROM, the diode ROM, that was used in a variety of applications before the development of integrated circuits. There are eight addresses, numbered 0 through 7, at the top of the figure in the block labeled 'decoder'. The address lines run vertically down the diagram. In each address there are four data positions, corresponding to the four data lines running horizontally across the figure. In some positions the data line is connected to the address line by a diode. In these positions, continuity exists between address line and data line. In other positions an open circuit exists. Here there is no continuity between address line and data line. By constructing a diode ROM with selected positions of continuity and discontinuity, a device is created with permanently programmed information in binary form.

Extracting that information from ROM requires selecting a specific address to read. This selection is made via the address bus, which is shown holding the address 001. Consequently, an electronic switch connects address 1. When this occurs, the address 1 line is connected to ground as indicated by the switch position at the top of the diagram. Each data line is floating at 5 volts. When address 1 is brought to ground, continuity from the 5-volt source to ground exists in three data positions, those with the diodes. Therefore, these data positions are taken to ground and 0 volts is present on those data lines. However, where continuity is broken, the data line still reads 5 volts. Therefore, the data bus reads the logic string 0001 in address 1.

ROMs are used for permanent storage of information, e.g. machine operating instructions or language interpreters. They are programmed at the factory by starting with a device that has continuity at all intersections and destroying connections at address/data bit positions in order to establish the desired code and sequence of 0s and 1s.

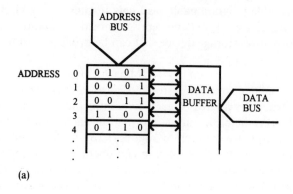

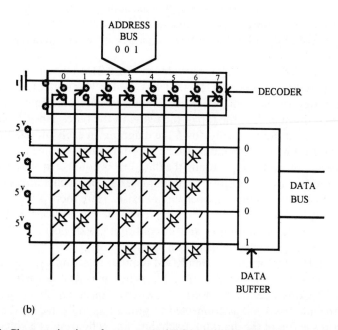

Fig. 2.21 Characterization of memory and ROM information storage and data transfer. (a) Data storage in memory addresses; (b) memory address decoding and data transfer.

An important device in the addressing scheme is the memory address decoder. A decoder is a logical device that can select a given output based on the number of unique boolean combinations at its input. For example, a decoder with one input is capable of selecting either of two outputs. A logical 1 at the input results in selecting one of the two outputs; a logical 0 at the input results in selecting the other output. Similarly a decoder with eight input lines can select $2^8 = 256$ unique outputs. Therefore, in order to access the eight

ARCHITECTURE AND OPERATIONS OF MICROPROCESSOR SYSTEM

addresses of Fig. 2.21, an encoder with three input address lines ($2^3 = 8$) is required. The gate structure (circuit) for the decoder of Fig. 2.21 can be constructed from what was learned in the previous section.

EXAMPLE 2.2

Design a decoder to select one of eight outputs from three inputs, as shown in Fig. 2.21.

Answer

To uniquely identify one of eight addresses using binary inputs requires only three binary digits as follows:

Code (input)	Corresponding address (output)
000	Address line 1
001	Address line 2
010	Address line 3
011	Address line 4
100	Address line 5
101	Address line 6
110	Address line 7
111	Address line 8

To design the digital circuit corresponding to this set of input/output relationships, we follow the procedure of section 2.5.

Step 1: Construct the truth table.

Inputs			Outputs							
A	B	C	X_1	X_2	X_3	X_4	X_5	X_6	X_7	X_8
0	0	0	1	0	0	0	0	0	0	0
0	0	1	0	1	0	0	0	0	0	0
0	1	0	0	0	1	0	0	0	0	0
0	1	1	0	0	0	1	0	0	0	0
1	0	0	0	0	0	0	1	0	0	0
1	0	1	0	0	0	0	0	1	0	0
1	1	0	0	0	0	0	0	0	1	0
1	1	1	0	0	0	0	0	0	0	1

Step 2: Form the logical products of the outputs that are true.

$$X_1 = \bar{A} \cdot \bar{B} \cdot \bar{C}$$
$$X_2 = \bar{A} \cdot \bar{B} \cdot C$$
$$X_3 = \bar{A} \cdot B \cdot \bar{C}$$
$$X_4 = \bar{A} \cdot B \cdot C$$
$$X_5 = A \cdot \bar{B} \cdot \bar{C}$$
$$X_6 = A \cdot \bar{B} \cdot C$$
$$X_7 = A \cdot B \cdot \bar{C}$$
$$X_8 = A \cdot B \cdot C$$

Step 3: Since there is only one true state for each output, the logical sum is not required.

The digital circuit implementation of a one-of-eight decoder can now be drawn as shown in Fig. 2.22.

An important observation to make from Example 2.2 is that the number of unique addresses that can be enabled by a decoder is a function of the number of binary combinations on the input. In general, for an address bus of N lines, there are 2^N addresses that can be accessed.

The address bus is used to address all memory locations, both ROM and RAM. RAM is volatile memory, i.e. its current state can be altered by writing

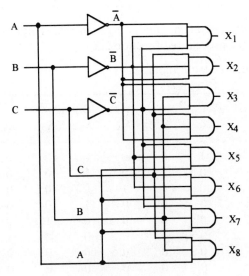

Fig. 2.22 A 1-of-8 decoder.

ARCHITECTURE AND OPERATIONS OF MICROPROCESSOR SYSTEM

to it or by turning the power off. RAM is constructed from bistable electronic circuits called flip flops. The output of a flip flop holds the binary data programmed or read into RAM. The output remains stable until the input combinations are changed. We will give some insight into the structure of RAM by examining the design of flip flops and address registers in section 2.6.2.

Referring again to Fig. 2.20, we shall now discuss some of the important components of the microprocessor. These components are often treated as though they were on the microprocessor chip. This may or may not be true depending on a particular computer architecture.

The **control memory** is shown in the upper right-hand corner of Fig. 2.20. Its purpose is to control the various parts of the microcomputer. The **arithmetic logic unit**, shown at the lower right, is a set of combinatorial networks that perform arithmetic and logical operations on data.

The computer's normal operating function is to read instructions from memory, to execute those instructions, and to write results back to memory. This is known as the 'machine cycle'. Several components of the microprocessor are instrumental in executing that cycle. We will discuss these components beginning with the program counter.

The **program counter** is a register that holds the binary address of the next instruction that will be fetched from memory. The **memory address register** holds the contents of the current address being accessed in memory. Therefore, there is a sequence wherein the program counter passess an address to the memory address register. The memory address register is physically connected to the address bus. The contents of the memory address register define the address on the address bus. This address is decoded and accessed as previously described.

The **memory data register**, shown at the upper left, takes the information via the data bus from the address that is currently being accessed (read mode) or writes information to that address (write mode). Information coming through the memory data register can be of two types: instructions and data. Instructions are binary strings that define a particular operation to be performed by the microprocessor. Data are binary strings to be operated on in accordance with instructions. When instructions are brought into the memory data register, they are passed to the **instruction register**. The instruction register holds the instruction while it is decoded and executed. When data is brought into the memory data register, it is passed to the **accumulator** or other arithmetic register. The accumulator is a kind of scratch pad for arithmetic and logical operations.

The registers described above, the program counter, memory address register, memory data register, instruction register, and accumulator, execute a repetitive cycle of events which will be described in some detail in the next section. Sometimes that cycle of events is broken or interrupted in order for the machine to perform some other function. For example, a controller may be performing the task of repetitively collecting sensor information at an input and storing that information in memory. An operator may be allowed to interrupt this process in order to request that the history of data be printed out. In such an event, the computer must retain information on where it left off before it was diverted to attend to this new request. The internal stack supports this requirement.

The **internal stack** is memory used by the microprocessor to temporarily store information. In the case just described, the current contents of the program counter, instruction register, memory address register, and accumulator would be stored by the microprocessor before it began to service the operator's request. When the request is completed, the stored information can be reloaded into these registers and the machine can begin to function where it had left off. The **stack pointer** is a register that holds the highest address of the stack in which information has been stored. This is a marker that informs the microprocessor of how far into the stack data has been placed.

This background will assist in explaining the operating cycle of the computer, which is the subject of the next section.

2.6.1 The machine cycle

Figure 2.23 shows a simplified hypothetical bus structure connecting a microprocessor with memory. There are three address bus lines which, after

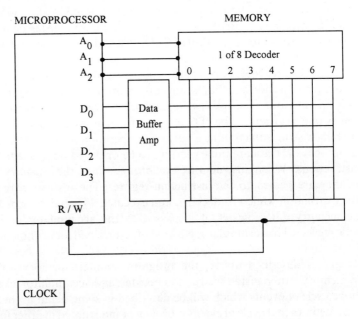

Fig. 2.23 Hypothetical bus structure for memory read/write operations.

ARCHITECTURE AND OPERATIONS OF MICROPROCESSOR SYSTEM

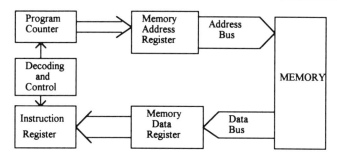

Fig. 2.24 Typical information flow during machine cycle.

decoding, enable eight memory locations (addresses). There is a four-bit data bus for reading and writing to memory. Finally, there are control lines. The read/write (R/W) line is the control line that indicates whether a read or write operation is to take place. When the signal on this line is low (logic 0), data will be written to memory; when the line is high (logic 1), data will be read from memory. Another control signal is the clock. In the design of a microprocessor-based system it is necessary to have a way of coordinating the order and the timing of execution. This is done with a square wave generator called the clock. The clock provides evenly spaced square pulses. At each pulse, some portion of an instruction is executed; an instruction consists of one or more clock cycles. The speed at which the clock operates in part accounts for the speed of execution of instructions.

The operation of a microcomputer can be described in terms of a sequence of interrelated cycles. The clock cycle is the smallest of the cycles and consists of one clock pulse. The 'machine cycle' is the longest cycle and consists of two subcycles: the fetch cycle and the execute cycle. During the fetch cycle data is acquired (fetched) from memory and brought into the microprocessor. During the execute cycle data is operated on within the microprocessor.

Figure 2.24 illustrates the sequence of information flow during a machine cycle. During the fetch cycle the following events take place:

- The program counter loads into the memory address register the address of the next memory location from which a programmed instruction is to be read.
- The address is put on the address bus, decoded, and the address is enabled.
- The contents of the address are read into the memory data register of the microprocessor.

During the execute cycle the instruction is decoded and the requested operation is performed. The following events take place:

- The contents of the memory data register are transferred to the instruction register of the CPU.

- The instruction is decoded by the instruction decoder.
- The instruction is executed.
- The program counter is incremented and the next fetch cycle begins.

An example of this cycle can be illustrated using a typical machine-level instruction. The hexadecimal code B6 is a machine language instruction for the Motorola 6800 microprocessor. B6 means 'load the accumulator'. As previously described, the accumulator is a register in the microprocessor that holds data while it is being operated on. It is a sort of scratch pad. The B6 instruction is followed by a hexadecimal argument that is a 16-bit number. That 16-bit number is the address in which the data that is to be placed in the accumulator resides. The instruction 'B6 1027' instructs the microprocessor to load the accumulator with the contents of address 1027.

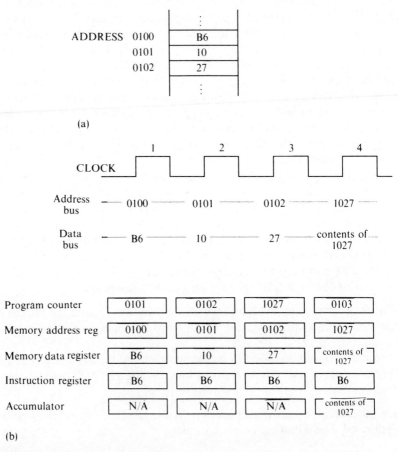

Fig. 2.25 Execution of instruction B6 in four clock cycles. (a) Instruction residing in sequential memory addresses; (b) data transfer during clock cycles.

ARCHITECTURE AND OPERATIONS OF MICROPROCESSOR SYSTEM

A user-written machine language program is stored in sequential memory locations in RAM. Figure 2.25a shows a user-written program with instruction B6 in address 0100 and the hexadecimal argument in addresses 0101 and 0102.

Figure 2.25b shows the execution of this instruction in relation to the clock cycle. This instruction executes in four clock cycles. On the first clock pulse, the contents of the program counter is placed in the memory address register, 0100 appears on the address bus and B6 is fetched from 0100 via the data bus. The instruction goes into the instruction register. The next two clock cycles result in fetching the address of the data, which is placed into the program counter. The B6 instruction has been fetched and it is now to be executed.

On the fourth clock cycle, the address 1027 is placed in the memory address register (thus, on the address bus) and the contents of 1027 are acquired and placed into the accumulator. Figure 2.25b shows the contents of the key registers during the execution of each clock cycle.

Referring again to Fig. 2.23, it is interesting to examine the sequence of events that occurs within the clock cycle. For the hypothetical system described, this activity is shown for a memory read operation using the timing diagram of Fig. 2.26. The cycle begins when a valid address is placed on the address bus; for example, the address 1027 of the fourth clock cycle of Fig. 2.25b. For the read operation, the $R/\overline{W}$ line is taken high (logic 1). After a time delay, the CPU reads the data on the data bus. For typical IC devices, this all happens within the microsecond range. Figure 2.26 is a simplified illustration; accessing memory devices in actual systems typically requires additional control logic.

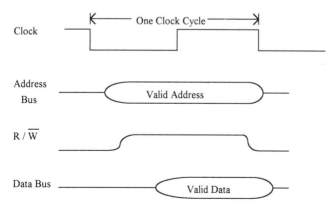

Fig. 2.26 Timing diagram for writing to memory.

2.6.2 Flip flops and registers

During the machine cycle, data or instructions are moved between devices. Data or instructions can be read from an address location (address register) to the microprocessor's memory data register. Some instructions require data to be transferred (written) from the microprocessor to an address register. The movement of data from and to a register can be illustrated using a sequential device called a **flip flop**. In this section we shall discuss flip flops and their application in the construction of registers.

Digital elements in a microcomputer (other than ROM) are constructed so as to be able to change state (1 to 0, 0 to 1) on command. For that reason they are called bistable (two-state) devices. The basic family of bistable devices is known as flip flops.

A flip flop is a 1-bit memory device for storage of a logic variable. It differs from a gate because it is a sequential device. This means that the current state of its output depends on the current state of its input and on the past history of the device. Typical sequential networks in which you will find flip flops are counters, shift registers, and memory registers.

One example of a widely used flip flop is the JK flip flop, which is illustrated in Fig. 2.27. Figure 2.27a shows the logic circuit, which consists of four AND gates and two NOR gates. The inputs are J and K, outputs are Q and $\bar{Q}$. Sequential networks are synchronized by the clock cycle, which is also shown on the input side.

When the device is enabled and a clock pulse occurs on the clock input at time t, the binary values on inputs J and K at time t and the current values of Q and $\bar{Q}$ will determine the output values on Q and $\bar{Q}$ at time $t+1$. This is shown in Fig. 2.27b. Figure 2.27b illustrates the response of the flip flop when the inputs are $J=0$, $K=1$. Each line of Fig. 2.27b represents an event in the sequence of operations. The letters A through F correspond to the logic gate at which the event takes place. Since $\bar{Q}$ is tied back to the input of AND gate A, the operation $0 \cdot \bar{Q} = 0$ is performed at A. The result, 0, becomes an input at C. Similarly, Q is the result of the operation at B. The clock input is '1'. Flip flops are triggered (change state) in synchronization with the leading edge of the clock. Therefore, at AND gate C, $0 \cdot 1 = 0$. The other operations work sequentially from the previous operations. Note the first pass at E, where NOR $(0 + Q) = \bar{Q}$. This becomes an input at F, where the next sequential operation results in 0. This result now becomes an input at E, which results in an output of 1. This becomes an input at F, which results in an output at F of 0. Once the output repeats itself, the output of the device is stable. Before the output is stable, the output of the device may flip and flop between values, hence the name 'flip flop'. It is left to the reader to trace the sequence of events from other initial conditions. Figure 2.27c shows the IEEE symbol for the JK flip flop.

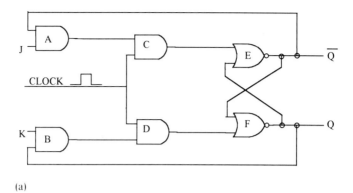

(a)

@ $A: 0 \cdot \bar{Q} = 0 \Rightarrow C$
@ $B: 1 \cdot Q = Q \Rightarrow D$
@ $C: 0 \cdot 1 = 0 \Rightarrow E$
@ $D: Q \cdot 1 = Q \Rightarrow F$
@ $E: \overline{0 + Q} = \bar{Q} \Rightarrow F$
@ $F: \overline{Q + \bar{Q}} = 0 \Rightarrow E$
@ $E: \overline{0 + 0} = 1 \Rightarrow \bar{Q}$
@ $F: \overline{1 + Q} = 0 \Rightarrow Q$

(b)

(c)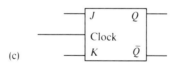

Fig. 2.27 Circuit and symbol for JK flip flop. (a) Logic network; (b) response of JK flip flop for $J = 0$, $K = 1$ and initial state $\bar{Q} = 0$, $Q = 1$; (c) IEEE symbol.

ARCHITECTURE OF MICROPROCESSOR-BASED SYSTEMS

The truth table for a sequential network, such as a flip flop, is defined for two sequential points in time and is given below. Time t_n is the period before the clock pulse; time t_{n+1} is after the clock pulse.

t_n		t_{n+1}	
J	K	Q	$\bar{Q}$
0	0	Q_n	$\bar{Q}_n$
0	1	0	1
1	0	1	0
1	1	$\bar{Q}_n$	Q_n

A flip flop commonly used to retain information in RAM is the D flip flop (data flip flop). As in the case of the JK flip flop, it is a clock-triggered device. In the D flip flop there is only one input and one significant output. The symbol for a D flip flop is given in Fig. 2.28 and its truth table appears below.

t_n	t_{n+1}	
D	Q	$\bar{Q}$
0	0	1
1	1	0

We will use the D flip flop to illustrate the construction and operation of a register, in this case a four-bit data register in RAM. This device is shown in two sequential states in Fig. 2.29.

Each D flip flop retains one state at its output (Q). Hence, if you cascade four D flip flops, you have a device that is capable of retaining four bits of information. Several commonly used RAM chips are constructed of registers which retain four bits of information.

Figure 2.29 illustrates a write operation. The data D_3–D_0 are four bits on the data bus to be written into this register (address), as shown in Fig. 2.29a. As explained in the previous section, during the machine cycle the address is

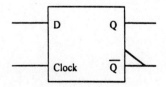

Fig. 2.28 Symbol for D flip flop.

ARCHITECTURE AND OPERATIONS OF MICROPROCESSOR SYSTEM

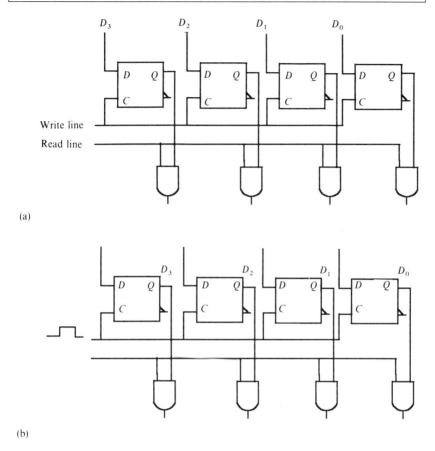

Fig. 2.29 Write operation in a 4-bit register. (a) Data on data bus at time m; (b) Data in storage at time $m + 1$.

placed on the address bus and a decoder selects the correct address register. The control circuitry then connects the write line to the clock for one clock cycle. When the clock pulses (traverses from logic 0 to 1 to 0), whatever is on the input side of the D flip flop register appears on the output, as shown in Fig. 2.29b. This data will remain in RAM until new data is presented and loaded.

Figure 2.30 illustrates the read operation. When data is to be read from RAM the control bus enables the read line by connecting it to the clock. When the clock pulses, whatever is in memory becomes the output of the AND gate, which is placed on the data bus for transfer to the microprocessor.

Another useful register to illustrate the application of a flip flop is a serial to parallel register. In a later section we will be discussing input/output devices. Data from outside a computer is often transferred into the computer one bit at a time through a device called a serial port. As we have seen in our discussion of

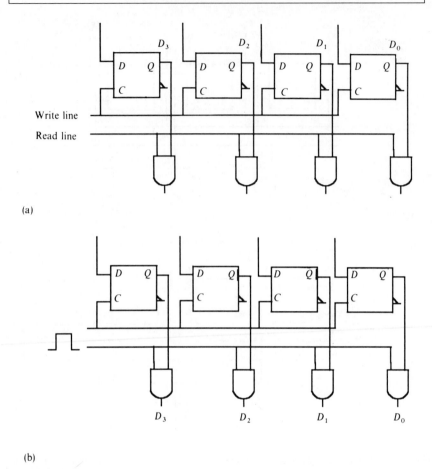

Fig. 2.30 Read operation from a 4-bit register. (a) Data in storage at time m, (b) Data on data bus at time $m + 1$.

the data bus, internally data is transferred several bits at a time, or in parallel. A serial to parallel converter accepts data in serial form and transfers it internally in parallel form.

Figure 2.31 illustrates the design and sequence of operations of a register that is loaded serially, or one bit at a time. In addition to the usual input and output lines, there is a reset line that clears the stored values in the flip flops. Initially the register is cleared using the reset line. On the first clock cycle the first bit in the data stream is stored in Q_3. The second clock cycle moves Q_3 to Q_2, and Q_3 becomes the second bit in the data stream. This continues until all four bits are loaded. At that point the control bus can enable the parallel data lines, as was previously shown in Fig. 2.30, and the next clock cycle transfers data along the data bus in parallel.

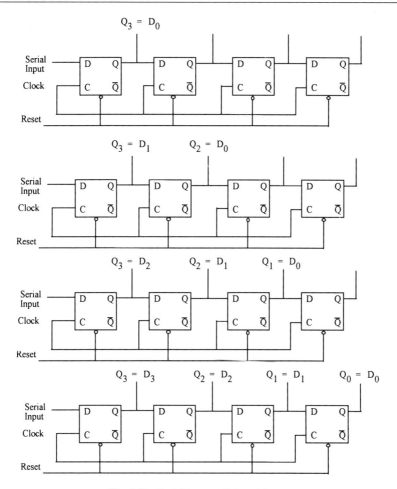

Fig. 2.31 Serial to parallel register.

2.7 NUMBER SYSTEMS AND LOW-LEVEL COMPUTER LANGUAGES

In our usual experience with mathematics we are used to thinking about numbers as base 10 symbols, i.e. the decimal system. From our previous discussion it should be clear that the natural number system of a digital circuit is based on two states (0, 1; false, true). This is the binary number system.

2.7.1 Binary number system

Any number system consists of a base and a set of symbols called digits. The integers of a number system can be written as follows:

$$d_n b^n + d_{n-1} b^{n-1} + \cdots + d_2 b^2 + d_1 b^1 + d_0 b^0,$$

where:

b is the base of the number system
d_n is the nth digit ($0 \leqslant d \leqslant (b-1)$)

Thus, in the decimal system, the number 622_{10} can be written:

$$6 \times 10^2 + 2 \times 10^1 + 2 \times 10^0 = 622$$

In the binary number system, $b = 2$ and $d = 0, 1$. Hence, the binary number 101_2 has components as follows:

$$1 \times 2^2 + 0 \times 2^1 + 1 \times 2^0 = 101$$

The relationship between decimal and binary numbers are given in Table 2.1 for some selected examples. This representation makes binary to decimal conversion straightforward. Taking the components of the number 1010_2

$$
\begin{array}{rl}
1 \times 2^3 = & 8 \\
0 \times 2^2 = & 0 \\
1 \times 2^1 = & 2 \\
0 \times 2^0 = & 0 \\
\hline
& 10
\end{array}
$$

Table 2.1 Relationship between decimal and binary numbers

Decimal Number	Decimal Components	Equivalent Binary Number	Binary Components
0	0×10^0	0	0×2^0
1	1×10^0	1	1×2^0
2	2×10^0	10	$1 \times 2^1 + 0 \times 2^0$
3	3×10^0	11	$1 \times 2^1 + 1 \times 2^0$
4	4×10^0	100	$1 \times 2^2 + 0 \times 2^1 + 0 \times 2^0$
5	5×10^0	101	$1 \times 2^2 + 0 \times 2^1 + 1 \times 2^0$
10	$1 \times 10^1 + 0 \times 10^0$	1010	$1 \times 2^3 + 0 \times 2^2 + 1 \times 2^1 + 0 \times 2^0$

NUMBER SYSTEMS AND LOW-LEVEL COMPUTER LANGUAGES

Decimal to binary conversion requires the inverse operation. Returning to the general format of a number system, we note:

$$d_n b^n + d_{n-1} b^{n-1} + \cdots + d_1 b^1 + d_0 = ((d_n b + d_{n-1})b \ldots)b + d_1 b + d_0$$

For example:

$$1 \times 2^3 + 0 \times 2^2 + 1 \times 2^1 + 0 = ((1 \times 2 + 0) \times 2 + 1) \times 2 + 0$$

To find the binary equivalent of a decimal number we note:

$$\frac{(((d_n b + d_{n-1})b \ldots)b + d_1)b + d_0}{b} = ((d_n b + d_{n-1})b \ldots)b + d_1 + \frac{d_0}{b}$$

Division by b is the inverse of successive multiplication by b. It yields a quotient:

$$((d_n b + d_{n-1})b \ldots)b + d_1$$

and a remainder:

$$\frac{d_0}{b}$$

The numerator of the remainder is d_0, the least significant digit of the binary number. Hence, successive division by b ($b = 2$) of a decimal number will yield a binary number. For the binary equivalent of 10_{10}:

```
2 | 10      with 0 remainder (least significant digit)
2 |  5      with 1 remainder
2 |  2      with 0 remainder
2 |  1      with 1 remainder (most significant digit)
     0
```

$$10_{10} = 1010_2$$

Arithmetic calculations are carried out on a digital computer using the laws of binary addition. Addition of binary digits obeys the following rules:

$$0 + 0 = 0$$
$$0 + 1 = 1$$
$$1 + 1 = 0 \quad \text{(carry 1)}$$
$$1 + 1 + 1 = 1 \quad \text{(carry 1)}$$

These rules are analogous to those used in decimal addition, where a carry occurs whenever the sum of two digits exceeds 9, the largest digit of the decimal system. Consider the addition of 1110 and 1100:

$$\begin{array}{r} 1\ 1\ 1\ 0\quad X \\ \underline{1\ 1\ 0\ 0}\quad Y \\ 1\ 1\ 0\ 1\ 0\quad S = X + Y \\ 1\ 1\qquad\quad \text{carry bits, } C \end{array}$$

When addition is requested by a program instruction, the microprocessor carries it out efficiently using a combinational network called an adding network. The duality between digital circuits and binary arithmetic is nicely illustrated by a circuit that performs the addition of two binary digits (called a half adder). From our previous description of constructing simple circuits from gates, we will derive the network for a half adder from its truth table.

EXAMPLE 2.3

Design a digital network that obeys the laws of binary addition as given above; i.e. it adds bits X_1 and Y_1 to obtain the sum (S_1) and a carry bit (C_1).

Answer

Step 1: Write the truth table.

Inputs		Outputs	
X_1	Y_1	S_1	C_1
0	0	0	0
0	1	1	0
1	0	1	0
1	1	0	1

Step 2: Form the logical product of the inputs for which the output is 1.

Output of S_1 Output of C_1:

$\overline{X_1} \cdot Y_1$ $X_1 \cdot Y_1$

$X_1 \cdot \overline{Y_1}$

Step 3: For each output, take the logical sum of the logical products.

$$S_1 = (\overline{X_1} \cdot Y_1) + (X_1 \cdot \overline{Y_1})$$
$$C_1 = X_1 \cdot Y_1$$

The digital network representation is shown in Fig. 2.32a.

NUMBER SYSTEMS AND LOW-LEVEL COMPUTER LANGUAGES

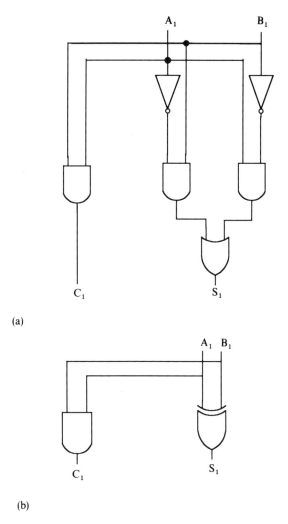

Fig. 2.32 Logical network for half adder. (a) Half adder; (b) half adder with EOR gate.

Note that the circuit that yields S_1 is an EOR gate, as previously defined in Fig. 2.19. Hence, Fig. 2.32a can be redrawn as shown in Fig. 2.32b.

Adders can be designed for more bits by simply cascading half adders. For example, the truth table for adding the second bits of two binary numbers would be as follows:

Inputs			Outputs	
X_2	Y_2	C_1	S_2	C_2
0	0	0	0	0
0	0	1	1	0
0	1	0	1	0
0	1	1	0	1
1	0	0	1	0
1	0	1	0	1
1	1	0	0	1
1	1	1	1	1

It is left as an exercise for the reader to apply the stepwise procedure to derive the circuit from the truth table.

2.7.2 2's complement arithmetic

The rules for subtracting binary numbers are as follows:

$$0 - 0 = 0$$
$$1 - 0 = 1$$
$$1 - 1 = 0$$
$$0 - 1 = 1 \quad \text{(borrow 1)}$$

Consider the subtraction of 1010 from 1110:

$$\begin{array}{ll} 1\ 1\ 1\ 0 & M \\ 1\ 0\ 1\ 0 & S \\ \hline 0\ 1\ 0\ 0 & D = M - S \end{array}$$

Microcomputers generally perform subtraction operations using addition. The most widely used method is called **2's complement arithmetic**, which is a special format of binary numbers in which the leftmost position is reserved for the sign of the number. A 0 represents the plus (+) sign and a 1 represents the minus (−) sign. The logic of 2's complement arithmetic can be explained by considering the subtraction operation, above. There are three variables involved in subtraction, the minuend (M), the subtrahend (S) and the difference (D). The subtraction operation is defined as $D = M - S$. Now, subtraction can also be defined as the addition of two numbers as follows:

$$D = M + (-S).$$

NUMBER SYSTEMS AND LOW-LEVEL COMPUTER LANGUAGES

Therefore, you can also obtain the difference by taking the negative of the subtrahend and adding it to the minuend. This is precisely what the computer does. It takes the negative of the binary number that is the subtrahend and adds it to the binary minuend.

This raises the question: 'What is the negative of a binary number?' By definition, the negative of any number is a number that, when added to itself, yields a result of zero. It can be shown that the negative of a binary number is found by taking its complement and adding 1. For example, consider the 4-bit binary number 0 0 1 1. This is 3 in decimal. The stepwise procedure for obtaining its negative value is:

$$\begin{array}{r} \text{number} = 0\ 0\ 1\ 1 \\ \text{complement} = 1\ 1\ 0\ 0 \\ +\ 1 \\ \hline 1\ 1\ 0\ 1 \end{array}$$

Adding the number and its negative value:

$$\begin{array}{rl} 0\ 0\ 1\ 1 & M \\ 1\ 1\ 0\ 1 & (-S) \\ \hline 1\ 0\ 0\ 0\ 0 & D = M + (-S) \end{array}$$

Note that the 4-bit number is zero. The addition has resulted in a 1 being carried over to the fifth place. In general, in order to obtain the sign and magnitude of the result of the computation, the sign is found in the leftmost bit and the magnitude is found by performing the inverse of the operation used to obtain the negative of a number, i.e. add 1 to the difference and then complement the result.

$$\begin{array}{rl} 1\ 0\ 0\ 0\ 0 & \leqslant \text{negative sign} \\ -\ 1 \\ \hline 1\ 1\ 1\ 1 & \leqslant D - 1 \\ \text{complement}\ 0\ 0\ 0\ 0 & \leqslant \text{magnitude zero} \end{array}$$

Since the magnitude is zero, the sign is not relevant.

EXAMPLE 2.4

Use 2's complement arithmetic to subtract 30_{10} from 14_{10}.

Answer

Plus 14 is 0 1 1 1 0 and +30 is 0 1 1 1 1 0. Therefore, −30 is Not (0 1 1 1 1 0) + 1, or 1 0 0 0 1 0. Performing the computation:

```
    0 1 1 1 0
  + 1 0 0 0 1 0
    ─────────
    1 1 0 0 0 0   ⩽ negative sign
          − 1
    ─────────
    1 0 1 1 1 1
    0 1 0 0 0 0   ⩽ magnitude = 16
```

The difference is a -16_{10}.

The use of 2's complement notation limits the magnitude of the number that can be held in a register to one bit less than the size of the register. Therefore, an 8-bit register will hold seven magnitude bits and one sign bit.

2.7.3 Binary coded decimal

Another common format for maintaining binary numbers in a register is the use of binary coded decimal (BCD). BCD uses 4-bit binary strings to represent the digits 0 to 9. Hence, the following relationships are defined in BCD as shown in Table 2.2.

Binary coded decimal is often used in circumstances where it is desirable to have a decimal output reading of a number residing in a register. There are special-purpose circuits available that will convert a binary code to a digital number on a numerical output display of a digital panel. For example,

Table 2.2 BCD and decimal equivalents

Binary string	Decimal equivalent
0 0 0 0	0
0 0 0 1	1
0 0 1 0	2
0 0 1 1	3
0 1 0 0	4
0 1 0 1	5
0 1 1 0	6
0 1 1 1	7
1 0 0 0	8
1 0 0 1	9
0 0 0 1 0 0 0 0	10
0 0 0 1 0 0 0 1	11
0 0 0 1 0 0 1 0	12
.	.
0 1 0 1 0 0 0 0	50
.	.
0 1 0 1 0 1 0 0 0 0 1 0	542

NUMBER SYSTEMS AND LOW-LEVEL COMPUTER LANGUAGES

a laboratory scale taking a weight measurement will often store the result in a register as a BCD number. That register will be wired to a digital output circuit that displays the weight as a decimal number. An exercise at the end of this chapter will familiarize the reader with such a display.

2.7.4 Hexadecimal number system

At the level of the electronic circuit, instructions are executed in binary. This is the natural representation from the hardware. For that reason, instructions coded in binary are usually referred to as machine language instructions. Several lines of code in machine language for a Motorola 6800 microprocessor are shown below. This is a program to add two numbers:

```
       PROGRAM
       1011  0110
       0000  0101
       0001  0000
       1011  1011
       0000  0101
       0001  0001
       1011  0111
       0000  1010
       0000  0000
```

In order to simplify the readability of machine code, it is useful to represent strings of binary digits using hexadecimal numbers. The hexadecimal number system is the base 16 number system. Consequently, there are 16 symbols (0 to 9 and A to F). The correspondence between the hexadecimal, decimal, and binary number systems is illustrated in Table 2.3. Note the parsimonious representation of binary strings by hexadecimal symbols. The 4-bit binary string $1\ 1\ 1\ 1_2$ is represented by a single digit F. This eases the burden of programming instructions and is the rationale for using hexadecimal representation in writing low-level code.

The conversion between hexadecimal and other number systems follows the same procedure we have examined earlier. For example, conversion from decimal to hexadecimal requires successive divisions by the base 16. Consider converting the number 1250_{10}.

16 ⌊1250   with remainder $2_{10} = 2_{16}$ (least significant digit)
16 ⌊78   with remainder $14_{10} = E_{16}$
16 ⌊4   with remainder $4_{10} = 4_{16}$ (most significant digit)
   ⌊0

$$1250_{10} = 4E2_{16}$$

Table 2.3 Equivalent Numbers for Decimal, Binary and Hexadecimal

Decimal	8-bit binary	Hexadecimal	Decimal	8-bit binary	Hexadecimal
0	0000 0000	0	16	0001 0000	10
1	0000 0001	1	17	0001 0001	11
2	0000 0010	2	18	0001 0010	12
3	0000 0011	3	19	0001 0011	13
4	0000 0100	4	20	0001 0100	14
5	0000 0101	5	21	0001 0101	15
6	0000 0110	6	22	0001 0110	16
7	0000 0111	7	23	0001 0111	17
8	0000 1000	8	24	0001 1000	18
9	0000 1001	9	25	0001 1001	19
10	0000 1010	A	26	0001 1010	1A
11	0000 1011	B	27	0001 1011	1B
12	0000 1100	C	28	0001 1100	1C
13	0000 1101	D	29	0001 1101	1D
14	0000 1110	E	30	0001 1110	1E
15	0000 1111	F	31	0001 1111	1F

Converting back to decimal we simply multiply the decimal digit by the base 16 raised to the power of the digit position. Consider, for example, the number $4E2_{16}$:

$$
\begin{array}{lrl}
4 & 4 \times 16^2 = & 1024 \\
E & 14 \times 16^1 = & 224 \\
2 & 2 \times 16^0 = & \underline{2} \\
& & 1250
\end{array}
$$

These methods can be employed to convert from binary to hex and vice versa. However, it is much easier to perform such conversions by inspection using Table 2.3. Note that each hex symbol represents a 4-bit binary number. Hence, any binary number can be partitioned into 4-bit strings and each 4-bit string can be replaced by its equivalent hex digit. Consider, for example, the following binary string:

$$1\ 1\ 0\ 1\ 0\ 1\ 1\ 0\ 1\ 1\ 0\ 0\ 1$$

Separating 4-bit strings from right to left:

$$0\ 0\ 0\ 1\quad 1\ 0\ 1\ 0\quad 1\ 1\ 0\ 1\quad 1\ 0\ 0\ 1,$$

which converts to:

$$1\quad A\quad D\quad 9.$$

$$1\ 1\ 0\ 1\ 0\ 1\ 1\ 0\ 1\ 1\ 0\ 0\ 1_2 = 1\ A\ D\ 9_{16}$$

Hexadecimal to binary conversion is just the inverse operation.

The machine language program that was illustrated earlier is converted to its hexadecimal equivalent below. This representation is often referred to as operation code, or simply, op code.

```
        PROGRAM
          B6
          05
          10
          BB
          05
          11
          B7
          0A
          00
```

2.8 ASSEMBLY LANGUAGE PROGRAMMING

Machine language instructions and their op code equivalent are microprocessor specific. The physical design of the device is what enables the microprocessor to convert the binary strings of 1s and 0s to the opening and closing of connections in the electronic circuit. Ultimately, all programs that run on a computer are converted to machine language before they are processed.

Programming is normally done in a higher-level language using symbols and words that are easily understood and usually suggest the functions they represent. Such languages as C, Pascal, and Fortran are important languages today in engineering and manufacturing environments.

Programming at the level of the machine is often useful because it allows the programmer to be very specific about how he or she wants the system to execute a command, and the execution time can be controlled precisely. This can be helpful in reducing the time of execution of a specific instruction. For that reason, programs written in high-level languages may have subroutines that the programmer has written in a language closer to the operational level of the machine. For example, this can be accomplished by coding instructions in op code. More likely, a programmer will use assembly language. Assembly language is just a mnemonic representation of op code and machine language, i.e. the instruction sounds like what it wishes to accomplish. Since there is a one-to-one correspondence between assembly language instructions and op code, the op code program shown in section 2.7 can be converted to an equivalent assembly language program. Like the op code, assembly language is microprocessor specific. Below is illustrated the assembly language version of the op code program for the Motorola 6800 series microprocessor, a popular 8-bit microprocessor.

PROGRAM

```
LDAA  0510
ADDA  0511
STAA  0A00
```

Table 2.4 Partial MC6800 Assembly Language Instruction Set

Mnemonic	Operation	Addressing Mode	Op Code	Clock Cycles
ADDA (addr)	Add contents of address to contents of accumulator A	Extended	BB	4
ANDA (addr)	AND contents of address with contents of accum A	Extended	B4	4
BEQ (rel)	Branch to (rel) if result of last operation equals zero	Relative	27	3
BGE (rel)	Branch to (rel) if result of last operation is $\geq$ zero	Relative	2C	3
BNE (rel)	Branch to (rel) if result of last operation not zero	Relative	26	3
DECA	Decrement contents of accumulator A by 1	Inherent	4A	2
DEX	Decrement contents of index register X by 1	Inherent	09	3
EORA (addr)	Exclusive OR contents of addr with accumulator A	Extended	B8	4
INCA	Increment contents of accumulator by 1	Inherent	4C	2
INCB	Increment contents of accumulator B by 1	Inherent	5C	2
INX	Increment contents of X register by 1	Inherent	08	3
JMP (addr)	Jump to address	Extended	7E	3
JSR (addr)	Jump to subroutine located at address	Extended	BD	6
LDAA (addr)	Load accumulator A with contents of address	Extended	B6	4
LDAA (num)	Load accumulator A with the number	Immediate	86	2
LDAB (addr)	Load accumulator B with contents of address	Extended	F6	4
LDX (num)	Load the X register with the number	Immediate	CE	3
RTI	Return from interrupt	Inherent	3B	12
RTS	Return from subroutine	Inherent	39	5
STAA (addr)	Store the contents of accumulator A in address	Extended	B7	4
STAB (addr)	Store the contents of accumulator B in address	Extended	F7	4

ASSEMBLY LANGUAGE PROGRAMMING

It is beyond the interest of this book to go into the details of a specific assembly language. However, some familiarity with how a microprocessor executes instructions is gained by looking at some examples of assembly code. We will use the simple assembly language program above to illustrate some assembly language instructions and their relationship to microcomputer architecture. For this discussion, a partial 6800 instruction set is given in Table 2.4.

Figure 2.33 is a simplified representation of the key internal registers of the 6800 microprocessor. We have previously discussed the role of the program counter and the stack pointer. The two accumulators are used as 'scratch pads' for performing operations on data. The first instruction encountered in our program in LDAA 0510. Literally this says: load accumulator A with the contents of memory location 0510_{16}. This instruction commands the micro-

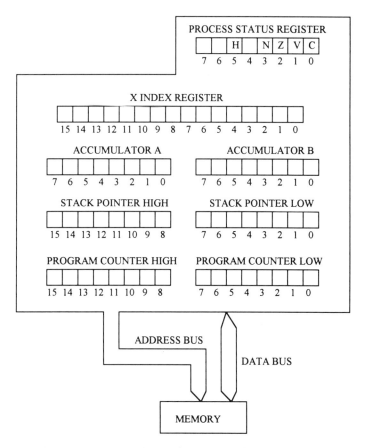

Fig. 2.33 Internal registers of the 68000 microprocessor.

processor to enable the memory location 0510 and read the data from the data bus into the accumulator.

The remaining instructions of the assembly language program and their interpretation are as follows:

Instruction	Description
ADDA 0511	Add the contents of memory location 0511_{16} to accumulator A. On this command the current contents of the accumulator (previously loaded from memory address 0510_{16}) and the contents of 0511_{16} are operated by adding circuit. The result will be an 8-bit binary number residing in the accumulator.
STA 0A00	Store the contents of the accumulator into memory location $0A00_{16}$. On this command, the 8-bit sum residing in the accumulator is placed in address $0A00_{16}$.

The example program simply took two bytes from two memory locations, added them together, and placed the resulting first eight bits into a new memory location. This example illustrates the manner in which the microprocessor interacts with peripheral chips, such as memory. Both the assembly language program and the data reside in memory. Execution of the program takes place in the microprocessor. Instructions are fetched from memory, decoded and executed within the microprocessor. The total execution time for an instruction is given in the last column of Table 2.4. For example, the LDAA instruction requires four clock cycles, the ADDA instruction requires four clock cycles, and so forth. By summing the execution time of each instruction, it is found that the program executes in 12 clock cycles. The conversion to seconds requires knowledge of the clock speed. Typical speeds are in the range of 1 to 25 MHz, yielding a clock cycle in the microsecond range.

2.9 COMMUNICATIONS AND AUTOMATIC DATA TRANSFER

In order for a computer or controller to be useful in automating a manufacturing system it must be able to communicate with external devices. These external devices may be sensors for monitoring a process, actuators for controlling a process, or the external device may be another computer. This and the following sections are concerned with the mechanics of how data and instructions are transferred in and out of a microcomputer or industrial controller. We will examine the organization of the hardware involved as well as the two primary methods of transferring binary data, parallel and serial transmission.

COMMUNICATIONS AND AUTOMATIC DATA TRANSFER

Information exchanges between devices requires the standardization of the language of communications. The American Standard Code for Information Exchange (ASCII) is such a language and it is used by most computer manufacturers. The ASCII code defines the relationship between alphabetic, numeric and punctuation characters and their corresponding binary string. The code is shown in Fig. 2.34. So, for example, to transmit the letter A from one machine to another, the transmitting machine would send the binary string 1000001. The receiving machine would be programmed to interpret the binary string as the letter A. ASCII provides binary symbols as well for control actions, such as carriage returns (CR) and line feed (LF).

ASCII is a 7-bit code. In serial communications, the eighth bit is often used as a check bit or parity bit to help insure the integrity of the transmission. The value of the eighth bit is specified by the parity convention, which has to be standardized between devices. Even parity requires that the sum of all 1s in the transmitted binary word be even. Odd parity requires that the sum be odd. Hence, transmitting the letter A with odd parity requires that the parity bit position be a 1. Parity will be discussed in a later section on serial data transmission.

Most significant 3 bits

		000	001	010	011	100	101	110	111
	0000	NUL	DLE	SP	0	@	P	`	p
	0001	SOH	DC1	!	1	A	Q	a	q
	0010	STX	DC2	"	2	B	R	b	r
	0011	ETX	DC3	#	3	C	S	c	s
	0100	EOT	DC4	$	4	D	T	d	t
	0101	ENQ	NAK	%	5	E	U	e	u
Least	0110	ACK	SYN	&	6	F	V	f	v
significant	0111	BEL	ETB	'	7	G	W	g	w
4 bits	1000	BS	CAN	(	8	H	X	h	x
	1001	HT	EM	)	9	I	Y	i	y
	1010	LF	SUB	*	:	J	Z	j	z
	1011	VT	ESC	+	;	K	[	k	{
	1100	FF	FS	,	<	L	\	l	|
	1101	CR	GS	-	=	M	]	m	}
	1110	SO	RS	.	>	N	^	n	~
	1111	SI	US	/	?	O	_	o	DEL

Fig. 2.34 ASCII code.

2.9.1 Interfacing hardware organization

There are basically two primary techniques of data transfer used in computer interfacing: programmed data transfer and direct memory access (DMA). Each technique utilizes a different set of hardware and software control protocols. In this section we examine programmed data transfer, which is the method most commonly used in manufacturing applications.

Figure 2.35 is a block diagram illustrating the logic of a programmed data transfer that assumes a memory mapped input/output (I/O). As the name implies, the passage of each word of the data transfer is controlled by programmed instruction. For example, a programmed instruction may instruct the microprocessor to transfer the current contents of a memory address to the peripheral device through its output port. This would be done in several steps. First, data is read to memory. Then, the contents of a memory register would be transferred to the output port. This data is then available externally on the port pins or other external connection that is electronically connected to the peripheral device. Typically the program also provides instructions by which the peripheral device is informed that data is available. The programmed data transfer may also provide instructions that coordinate the passing of data across the interface boundary using interface control lines. Programmed data transfer is commonly used in microcomputer applications and it will be the technique underlying most of our discussion of parallel and serial communication.

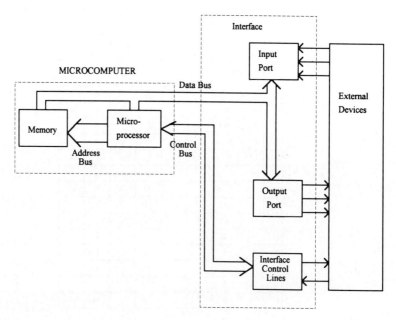

Fig. 2.35 Components of a computer system involved in a programmed data transfer.

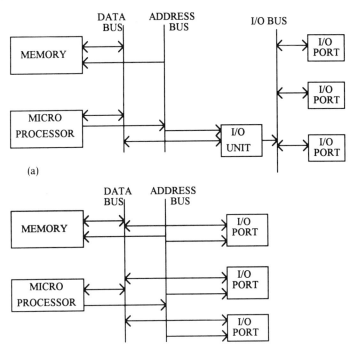

Fig. 2.36 Two implementations for port addressing. (a) Dedicated I/O; (b) memory mapped I/O.

There are differences in the input/output interface architecture of computers and controllers based on the bus implementation and the microprocessor used in the system. There are basically two classes of I/O bus implementation: dedicated I/O bus and memory mapped I/O. These differences are illustrated in Fig. 2.36.

Figure 2.36a is a block diagram of a dedicated I/O bus system. When an I/O device is addressed it is done so by using a port addressing command followed by a device number. The I/O unit interprets the instruction and selects the proper device for the data transfer. Dedicated I/O is a common implementation in the Intel family of microprocessors. Two assembly language instructions in the Intel 8080 instruction set are IN (device) and OUT (device), which have the function of transferring data in or out respectively, from the argument (device). Table 2.5 is an example of an assembly language routine for outputting the contents of a memory location to a peripheral device.

Figure 2.36b is a block diagram illustrating the implementation of memory mapped I/O. Memory mapped I/O treats an output port as though it were just another memory location. A separate I/O bus is not required since all data moves along the data bus and the port address space is accessed through the address bus. Refer to Fig. 2.37. This is a memory map for the Motorola

ARCHITECTURE OF MICROPROCESSOR-BASED SYSTEMS

Table 2.5 Typical I/O commands in a dedicated I/O system

8080 assembler code	Remarks
LDA 6A20	Load accumulator with contents of memory 6A20
OUT OPORT	Output the contents of the accumulator to output port
IN IPORT	Input to the accumulator the data at input device port
STA 0100	Store the contents of the accumulator in memory location 0100

68HC11EVBU controller, a popular 8-bit microcontroller that uses the 6800 assembly language instruction set. A memory map shows how computer memory is partitioned for different functions. Here we see that addresses 0100 to 01FF are RAM addresses available for user programming. Some of the address space is actually reserved for the I/O ports, as shown in Fig. 2.37. Any data written to or read from that space is actually written to or read from an I/O device. Systems based on the MOS 6502 microprocessor and

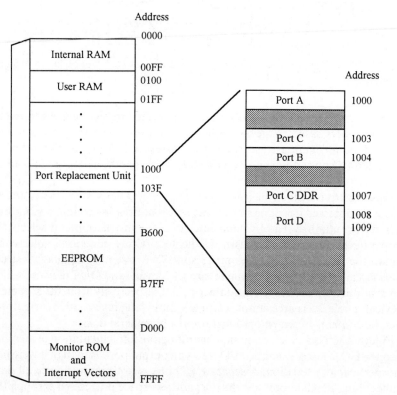

Fig. 2.37 Memory map of Motorola MC68HC11EVBU controller in signal chip mode.

COMMUNICATIONS AND AUTOMATIC DATA TRANSFER

Table 2.6 Typical I/O commands in a memory mapped I/O system

6800 assembler code	Remarks
LDAA 6A20	Load accumulator A with contents of memory address 6A20
STAA 1004	Output contents of accumulator A at the port B that is accessed at memory location 1004
LDAA 1003	Input to accumulator A the data at the input port C that is mapped to memory location 1003
STAA 0100	Store the contents of accumulator A in memory location 0100

the Motorola 6800 microprocessor are memory mapped and the usual commands for addressing memory are used to perform I/O operations. Table 2.6 is an example of a 6800 assembly language routine for performing the same operations as Table 2.5.

In the next two sections we will describe the fundamentals of parallel and serial communications. When examples are shown for illustration, we will be assuming the use of programmed data transfer techniques and memory mapped I/O.

2.9.2 Parallel data communications

Parallel data communications allows several bits to be transferred simultaneously across the interface between communicating devices. An example of two parallel ports are port B and port C of the microcontroller shown in the block diagram of Fig. 2.38, which is based on the memory mapping of the Motorola 68HC11EVBU. Each of these ports have eight parallel bits, indicated as 0 to 7. Port B is mapped to memory address 1004 and port C is mapped to memory address 1003. This is consistent with the memory map of Fig. 2.37.

Port B is an output-only port. This is indicated by the directional arrows. Memory address 1004 can be thought of as the programmer's window to port B. The programmer can manipulate bits in 1004, which will then be reflected at the physical pin connectors of the port. Fig. 2.38 show the port pins that are associated with bit positions in 1004.

When the programmer places a 1 in a bit position, it is reflected as a five-volt signal between the corresponding pin and ground. When the programmer places a 0 in a bit position, there is no voltage drop between the corresponding pin position and ground. So, for example, storing the bit string 0 0 1 1 0 1 0 1 in 1004 results in five volts appearing on pins 37, 38, 40 and 42.

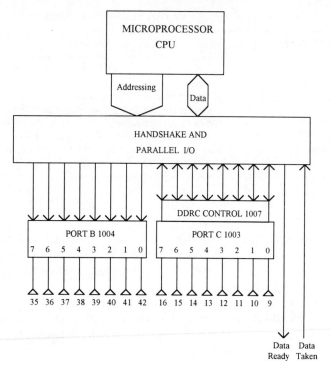

Fig. 2.38 Illustrated implementation of parallel ports.

EXAMPLE 2.4

Using the assembly language instructions of Table 2.4, write a program to output five volts each on pins 35, 37, 39, 40 and 41, zero volts otherwise.

Answer

Program	Op code	Comments
LDAA #AE	86 AE	Load accumulator A with the number AE
STAA 1004	B7 10 04	Store the number in port B

Whereas port B is an output-only port, port C is a bidirectional parallel port; it can be used for output and input. Since port C is bidirectional, the programmer has the additional responsibility of setting the direction of data flow. This is done by using the port C data direction register (DDRC), which is

COMMUNICATIONS AND AUTOMATIC DATA TRANSFER

located in address 1007, as indicated in Fig. 2.38 and the memory map of Fig. 2.37. DDRC is an 8-bit register. Each bit of the data direction register (address 1007) controls a bit of the data register (address 1003). When a 1 is placed in a bit position of DDRC, the corresponding bit of port C is set for output. When a 0 is placed in a bit position of DDRC, the corresponding bit position of port C is set for input. Figure 2.38 shows the pins associated with port C.

EXAMPLE 2.5

Assume that the controller is to collect data from an external device over port C and retransmit it out over port B. Write a program to accomplish this.

Answer

Program	Op code	Comments
LDAA #00	86 00	Load accumulator A with the number 00
STAA 1007	B7 10 07	Set all bits of port C as input
LDAA 1003	B6 10 03	Load data at port C into accumulator A
STAA 1004	B7 10 04	Write data to port B

The basic operation of a parallel port under software control is shown in Example 2.5. Since the data direction register of port C is programmable, the direction of data flow on any bit can be altered during program execution.

A problem with the transmission of Example 2.5 is the lack of coordination between the microcontroller and the peripheral device. If the peripheral device is a sensor or unintelligent dedicated device, this may not be a problem. However, if the peripheral device is an intelligent device, such as a computer or another controller, it may be necessary to coordinate the transmission. Initializing a port to output a message does not ensure that the peripheral device is ready to receive the message. Similarly, reading an input port does not ensure that the peripheral device has loaded a proper message to be received.

In order to gain positive control over the data transfer, a 'handshake' can be used. The purpose of the handshake is so the controller and the peripheral device can communicate the status of the transmission to each other. We will illustrate a programmed handshake in which the controller transmits ASCII words on the first seven bits of port C with the following agreed-upon protocol for the controller and peripheral device:

1. The transmission is initiated by taking bit 0 of port B high, which tells the intelligent peripheral device that the controller requests to send a transmission.

2. When the intelligent peripheral has taken the data, it acknowledges that event by taking bit 7 of port C high.
3. The controller then disables bit 0 of port B and waits for the peripheral to disable bit 7 of port C.
4. At this point the handshake is complete and the controller can assert a request to send another ASCII character.

This program is shown for the transfer of one ASCII character in Table 2.7. A useful operation in handling I/O is the masking operation, programmed in LOOP1 and LOOP2 of the table. In LOOP1, BEQ is a conditional branch based on the previous operation, which is a masking operation. When accumulator A is ANDed with hex #80, the bitwise AND will result in 0s in the first seven bit positions of the accumulator. The eighth bit, bit 7, will be ANDed with a 1 resulting in the value that was originally residing in the accumulator in bit 7. Bit 7 of port C is the only bit that is of interest at this stage in the program. If it is a 0, the data has not been taken by the peripheral device and the program should keep looping. If it is a 1, the data has been taken by the peripheral device and the program should exit the loop. In order to examine a selected bit of a register while blocking out all other bits, the programmer uses the masking operation.

Most computer and controller manufacturers implement additional lines on parallel ports for handshaking. Referring again to Fig. 2.38, we illustrate 'data ready' and 'data taken' lines as having been implemented as handshaking lines. The details of the implementation of these lines differs among computer and microcontroller manufacturers, based on the microprocessor and communication chips used in their systems. The reader should refer to specific applications to obtain more information on this subject.

Table 2.7 Example of a programmed handshake

Program				Comments
BEGIN	LDAA	#7F		Make bits 0–6 of Port C output and bit 7 input.
	STAA	1007		
	LDAA	#41		Place the letter A in Port C.
	STAA	1003		
	LDAA	#01		Assert a request to send by taking bit 0 of Port B high.
	STAA			
LOOP1	LDAA	1003		Load the contents of Port C and mask all but bit 7. If the result equals 0, continue to loop.
	ANDA	#80		
	BEQ	LOOP1		
	LDAA	#00		Remove the request to send.
	STAA	1004		
LOOP2	LDAA	#1003		Load the contents of Port C and mask all but bit 7. If the result equals 1, continue to loop.
	ANDA	#80		
	BNE	LOOP2		

Parallel data transmission is relatively fast (compared to serial transmission) because eight bits are being transmitted simultaneously. However, the reliability of parallel transmission will degrade as the distance between the computer and the peripheral device increases. This is because of timing differences in the unsynchronized speed of transmission and read operations performed on eight lines. Also, wiring parallel lines over a significant distance is expensive. As transmission distances increase, it is necessary to use serial communication, which is the subject of the next section.

2.9.3 Serial data communication

Serial data transmission, as the name implies, transmits data sequentially one bit at a time. If the transmitter and receiver have their clock synchronized, the data can be transferred with the least amount of handshaking required. However, it is more typical that devices do not have their timing coordinated and the transfer must be done using a well-defined handshaking standard. This latter condition is called **asynchronous serial communication** and is a type of point-to-point serial communication commonly found in manufacturing. Here we will focus on asynchronous serial communication.

Asynchronous serial communication has been around since the days of the Morse code. Asynchronous communication allowed individuals at each end of the transmission line to transmit at their own speed and still be able to understand the messages. For such a system to work in a computer environment there are two requirements: a communications protocol and an electronic signaling standard. The communications protocol is a agreed-upon code for interpreting the transmitted string of data. The electronic standard defines the hardware configuration and voltage levels as well as the handshake procedure. We will describe each of these subjects.

Depending on the manufacturer's design of the computer, controller or other device, the communications protocol may be established in hardware, by software, or both. It must be the same for each communicating device. Typical protocol parameters that must be set are: the baud rate, the start bit, the number of data bits, parity and the number of stop bits.

The rate of data transfer in bit/s is known as the **baud rate**. Typical baud rates are between 110 and 19 200 bit/s. The slower speeds are used for communication to mechanical devices, such as a printer; the higher speeds are used for communication just between electronic devices.

If you know the baud rate, you can compute the number of words per second that are being transmitted. A word is a block of bits that transmit one ASCII character. If it requires a string of 11 bits to transmit one ASCII character at 110 baud, the transmission rate is 10 word/s.

The transmission of an ASCII character will require more than the seven bits that describe the ASCII character because of additional bits required by the protocol. Figure 2.39 shows a sequence of bits transmitting the letter T.

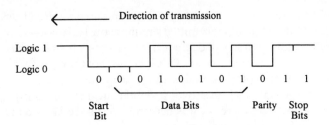

Fig. 2.39 Serial transmission of ASCII character T.

The character string must include a start bit to signal the beginning of a new data string. The start bit is 0. When the receiver encounters a start bit, it prepares to latch the next seven bits (the ASCII characters) into a serial register. This is then converted into parallel data to be transferred within the computer. Figure 2.31 showed an example of such a serial to parallel register.

Following the data bits there is an optional parity bit. The setting of a parity bit protocol allows the receiver to make an additional check on the validity of the transmission. The parity options are odd parity, even parity or no parity. If parity is enabled, the transmitter must ensure that the number of 1s in the data bits plus parity bit conform to the convention. For example, if odd partiy is chosen, and there are four 1s in the transmission of the seven data bits, a parity bit of 1 is required in order to conform to odd parity. If even parity is chosen, a parity bit of 0 is required. In Fig. 2.39, the letter T is being transmitted with odd parity.

Finally, the end of transmission of an ASCII character is indicated by one or two stop bits, depending on the agreed-upon convention. When these stop bits are encountered, the receiver knows that the next '0' bit is the start bit of a new ASCII transmission.

Once the protocol is set, the serial port hardware, called a UART (universal asynchronous receiver/transmitter), automatically conforms to the protocol, i.e. it is programmed to latch the correct bits into the data register and it provides an error signal when the parity check does not conform to the protocol. In addition to the protocol, each device must be using the same communications standard, or handshake, to determine which device is transmitting and which is receiving. Although there are different serial communication standards, the most widely used is RS232C.

The RS232 standard, issued by the Electronic Industries Association (EIA) in 1969, was written for the purpose of interfacing computer terminals to modems. Subsequently, the standard has been adopted by manufacturers of microcomputers, printers, machinery controllers and many others. This has, in fact, caused some difficulty for users since manufacturers often deviate from

COMMUNICATIONS AND AUTOMATIC DATA TRANSFER

strict adherence to the standard in designing their equipment. It is a great disappointment to a user to purchase a robot controller from one manufacturer and a vision system from another and then to find they cannot communicate, even though both have RS232 ports.

RS232 defines voltage levels and interface connections and their purpose. Logic level 0 is defined as a voltage level between $+3$ to $+15$ volts; logic level 1 is a voltage between -3 to -15 volts. The electronics at the interface of the communicating machines converts to and from transistor–transistor (TTL) logic level voltage (5 volts) which is used internally by the computer.

The standard defines a plug as the connector for a terminal and a receptacle for a modem. A terminal is data transmission equipment (DTE) and a modem is data communication equipment (DCE). Most implementations of the standard use either a 25-pin connector or a 9-pin connector. We will describe the functions of the transmission lines assuming a 25-pin connector.

The standard describes seven transmission lines and their function. Figure 2.40 shows five of these lines, which are the minimum configuration for most applications. Each pin is numbered and gives a descriptive word; the arrow indicates direction of the signal.

Pin 7 is circuit ground. When a DTE device and a DCE device are connected and turned on, each device signals their presence to the other. This is a power-up handshake and it is implemented on pins 6 and 20. Pin 6 indicates Data Set Ready (DSR). Pin 20 indicates Data Terminal Ready (DTR). The DTE device places a signal on pin 20 and DCE places a signal on pin 6. Before transmission begins, each device can test for the presence of the other. Software drivers written by the manufacturer of the computer or controller will often print a DEVICE NOT PRESENT error if the device is not on or not properly connected as indicated by the absence of a signal.

Pin 2 is known as the transmit line. It is the line over which transmitted data flows from DTE to DCE. Similarly, pin 3 is the receive line. It is the line over

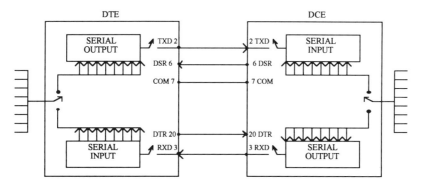

Fig. 2.40 Block diagram of five basic RS232 communication lines.

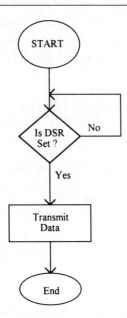

Fig. 2.41 Transmission control from DTE side using start-up handshake only.

which data is received by DTE from DCE. These lines take their name from the point of view of the DTE device.

Using this minimum configuration, the flowchart of Fig. 2.41 illustrates how the software control of a transmission from DTE to DCE might be accomplished. A check is made on the presence of DCE. A transmission occurs when DSR is set. It is assumed that DCE is a dedicated device and operates fast enough to keep up with the DTE transmission.

It is more typical that control is required during the transmission to prevent the data transmission rate from overflowing the buffers and to coordinate between devices so that only one device is trying to transmit at any time. This is the function of two additional status lines, pin 4 (request to send) and pin 5 (clear to send). These are shown in Fig. 2.42.

Request to send (RTS) is a DTE output, the purpose of which is to gain the DCE's attention and to inform it that data is ready to be transmitted. Clear to send (CTS) is DCE's response to that request. The CTS acknowledgment enables the UART transmitter at DTE and data is transmitted across the interface.

Figure 2.43 is a flow diagram of the activities on both the DTE and DCE side for the transmission of data. First, DTE confirms that DCE is on-line by checking line 6. It then submits a request to send over line 4. DCE is polling line 4 and responds by connecting its serial input register and acknowledging

COMMUNICATIONS AND AUTOMATIC DATA TRANSFER

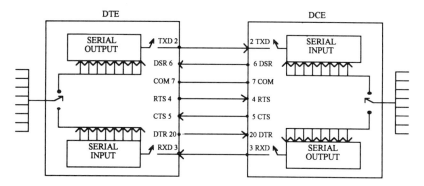

Fig. 2.42 Block diagram of seven RS232 communication lines.

by a signal on line 5. Data transmission then takes place. As the last word of the data transmission, DTE sends an end of transmission (EOT) character and disables line 4. DCE recognizes this as the end of transmission and disables line 5.

The RS232 standard and the conventions of DTE and DCE were set up for one purpose only: for terminals to communicate with modems. Microcomputers, industrial controllers, robot controllers, machine vision systems and so on often come equipped with a RS232 port and follow some or all of the RS232 signaling convention. However, it is not uncommon to purchase two pieces of equipment that must communicate with each other, both of which are configured as DTE. This kind of situation can be handled by reconnecting the cabling so that each device is fooled into thinking that a DCE device is attached at the other end. Figure 2.44 shows such an arrangement. Lines 20 and 6 are swapped at one end so that each device signals out at 20 and in at 6. Similarly, lines 2 and 3 are exchanged so that each device transmits on 2 and receives at 3. A similar arrangement is not logically possible on lines 4 and 5. Here we show the transmission handshake disabled by having the request line wired back to the CTS pin. Another resolution to initiating and terminating a transmission is to have an agreed-upon character passed between devices to open and close transmission.

Often manufacturers provide a switch on their controllers that allows the user to choose whether the device behaves as DTE or DCE. In many cases this does not completely resolve the problem because the manufacturer may go outside and standard in implementing the serial port operation. It is typical for a systems engineer interfacing communication between machines to require the manufacturers to provide detailed information on how the serial ports are implemented in order that the engineer can rewrite the communication software drivers. This is often the only way to ensure a common protocol between machines. In a later chapter we shall review current efforts directed

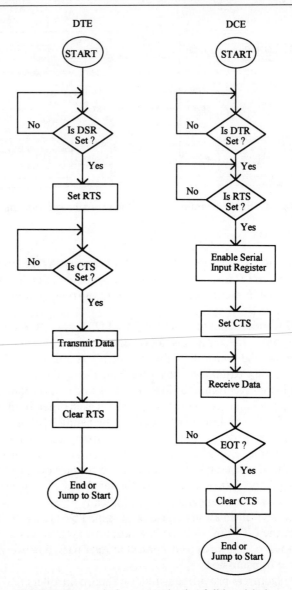

Fig. 2.43 Transmission control using full handshake.

toward solving the communications problem for equipment of different manufacturers. The integrity of data transmission has proven to be a headache in many automation projects and has led to the development of local area network protocols and standards that are supported by many manufacturers of controllers. We shall discuss this subject in Chapter 7.

SOFTWARE STRATEGIES FOR I/O COMMUNICATION

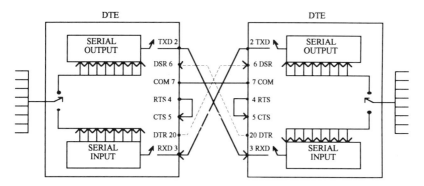

Fig. 2.44 Rewiring to enable two DTE devices to communicate.

2.10 SOFTWARE STRATEGIES FOR I/O COMMUNICATION

Input/output communication requires more than just setting the data direction register of the port and writing to it or reading from it. It is necessary to write software to organize the strategy for data collection. There are three basic software strategies for interfacing I/O:

1. continuous polling of I/O port,
2. periodic polling of I/O port, and
3. interrupt driven methods.

Continuous polling is useful, for example, in the case of a dedicated microcontroller. Here the controller is continuously monitoring a process in order to detect a change and to react to that change. Such a case is shown in Fig. 2.45a, where a temperature probe is monitoring a process on bit 0 of port C. If the temperature exceeds a certain limit, a signal is sent to the controller. When the signal is detected, the microcontroller turns on the fan. When the temperature returns below the critical point and the input on port C goes to 0, the fan is turned off.

In some applications it may be reasonable to sample the process intermittently, allowing for a specific time interval to occur between samples. **Periodic polling** is the name given to the strategy of sampling at specific intervals of time. A useful tool for periodic polling is the software timing loop, where a delay cycle in the program is predefined. Figure 2.46 is an example of a timing loop. This is accomplished by loading a register with a number and looping through a decrement instruction. The number n can be chosen by establishing the desired time delay and computing the number of loops that will achieve that delay. Loops can be nested within loops to yield longer time delays.

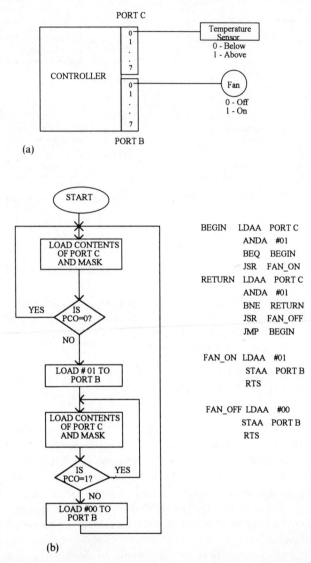

Fig. 2.45 Continuous polling. (a) Controller operating fan; (b) flow chart and program.

One obvious difficulty with the above examples is that the computer or controller is fully dedicated to a single task. It is often desirable to allow the computer to perform several functions and have it service a specific I/O function only when it is required to do so. In such cases the requirement can be initiated by allowing the peripheral device to interrupt the computer when service is required. This is the concept of an **interrupt driven system.**

SOFTWARE STRATEGIES FOR I/O COMMUNICATION

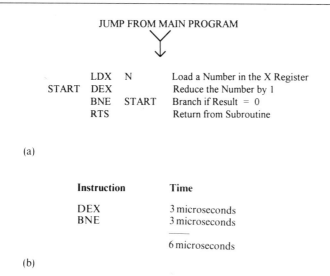

Fig. 2.46 Timing loop in periodic polling. (a) The timing loop; (b) time interval with 1-microsecond clock.

Microprocessors have pins called interrupt pins. When these pins are taken to ground, it causes the microprocessor to cease its current operation, store the contents of its registers in the stack, and jump to a predefined address to obtain instructions on what to do next. Figure 2.47 shows the pin assignments of the Motorola 6800 microprocessor; pins 4 and 6 are interrupts that can be wired to a peripheral device via the I/O port of a microcontroller.

We shall illustrate the concept of an interrupt using an IRQ interrupt as an example. The IRQ interrupt subroutine to handle external events on demand is written by the engineer designing the control application. The manufacturer of the computer or controller provides a set of memory locations to be used by the engineer for the application. When the IRQ pin is taken to ground, the microprocessor automatically stores the contents of its registers in the stack and jumps to a pair of addresses called the IRQ vector addresses. These are ROM addresses that are programmed by the computer or controller manufacturer at the factory. The content of these addresses is another address that the microprocessor loads into the program counter. In effect, the IRQ vector addresses are a pointer to another address. For example, the IRQ vector addresses for the MC68HC11EVBU are FFF2 and FFF3. As shown in Fig. 2.48, these memory locations contain the address 00EE.

The address 00EE is a RAM address. The engineer programs this address to direct the microprocessor to the user written subroutine for handling the

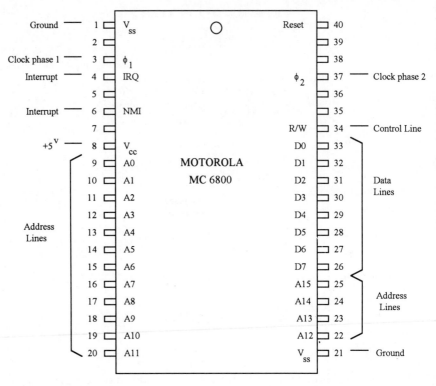

Fig. 2.47 Selected pins of the Motorola MC 6800 microprocessor.

interrupt. As shown in Fig. 2.48, 00EE contains the op code 7E, which is an unconditional JUMP statement (see Table 2.14). The next two addresses, 00EF and 00F0, contain the jump to location. Address 00EF contains the high byte to be loaded into the program counter (PCH) and 00F0 contains the low byte. On the next machine cycle, the microprocessor accesses address PCH–PCL. This is the address the engineer uses as the first address of the interrupt handling program. Naturally, it must be located in user RAM or programmed in EPROM. The user program must end with the assembler instruction RTI (return from interrupt). When this instruction is encountered, the microprocessor unloads the stack, placing its contents back into the registers as they were before the interrupt, and resumes the program that was being executed before the interrupt occurred.

SOFTWARE STRATEGIES FOR I/O COMMUNICATION

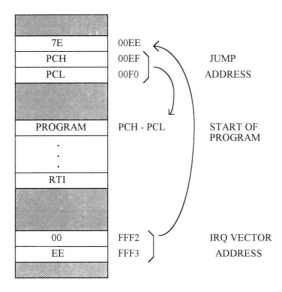

Fig. 2.48 Programming an IRQ interrupt.

EXAMPLE 2.6

Consider Fig. 2.49, which is a variation on the fan on/off control problem. Here bit 0 of port *C* is used for the sensory input and bit 7 of port *C* is used for the control output. Assume that the control program keeps track of the number of times the fan is turned on during the day. This is done by incrementing accumulator *B* each time the fan is turned on. Accumulator *B*, shown in Fig. 2.33, is an eight-bit register. An operator button has been wired to the IRQ interrupt. When the operator button is pressed, IRQ is brought to ground and the controller displays to the operator the current count in accumulator *B*. The eight bits of port *B* have been wired to a digital display for that purpose. The interrupt routine must write the current contents of accumulator *B* to port *B* (address 1004). Show the programming involved in setting up this interrupt.

Answer

The user must set the interrupt jump vector, starting in address 00EE and write the interrupt routine. One solution is as follows:

Jump addresses	Op code	Assembler
00EE	7E	JMP
00EF	01	01
00F0	00	00

Program Address	Op code	Assembler
0100	F7	STAB
0101	10	10
0102	04	04
0103	3B	RTI

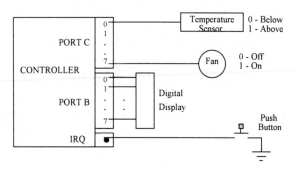

Fig. 2.49 Controller with IRQ pin wired externally.

2.11 SUMMARY

Microprocessor-based systems are made up of microprocessors, memory and input/output devices. The components are connected by an address bus, a data bus and appropriate control lines. The devices that make up the system are combinatorial and sequential logic devices. The simplest of these is the logic gate. By combining logic gates together, sequential logic devices can be created that are capable of temporarily storing and moving information. The natural language of these devices is binary. The programmed instruction for operating the computer written in binary is machine language instruction. For convenience, machine language instruction is often coded in hexadecimal format, which is called op code. For convenience of understanding and remembering, op code is given a mnemonic, which is called assembly language. High-level language instructions are made up of one or more assembly language instructions.

EXERCISES

In order for a computer or controller to be useful in an automated manufacturing system it must be able to communicate with other devices. Such communications may take the form of control actions or they may be the transmission of machine-readable data. When data is being transferred it is typical to use the ASCII data set which is a transmission standard for computer devices. The data may be transferred in parallel or serially, whichever is appropriate for the given situation. Each type of transmission has its own standards and protocol for controlling the transmission across the interface. These standards and protocols must be used by both communicating machines for the transmission to be understood.

The strategy for communications between computers or controllers and peripheral devices needs to be specified. In dedicated systems the strategy of continuous polling is often used. However, it is more common for computers and controllers to be assigned a number of tasks to control. Therefore, the use of interrupts is often required to initiate communication. An interrupt driven system requires user programming of an interrupt routine that handles the communication once the interrupt has occurred.

EXERCISES

1. An EXCLUSIVE NOR gate is an EXCLUSIVE OR gate followed by an inverter. What is its true table?
2. What is the truth table for the network in Fig. 2.50?

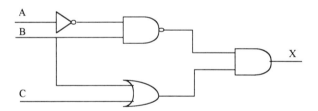

Fig. 2.50 Combinatorial network for Exercise 2.

3. Write the boolean expression for the following truth table, simplify it and give the minimum logic gate implementation.

A	B	X
0	0	1
0	1	0
1	0	1
1	1	1

4. Design a digital circuit for the following truth table.

Input		Output		
A	B	X	Y	Z
0	0	0	0	0
0	1	0	1	1
1	0	1	1	0
1	1	1	1	1

5. Using the properties of a boolean variable, simplify the answer to Exercise 4 as much as possible.
6. Simplify the following expressions and draw their simplest logic network.
   (a) $X = (\bar{A} \cdot \bar{B} \cdot C) + (A \cdot \bar{B} \cdot C) + (A \cdot B \cdot \bar{C}) + (A \cdot B \cdot C)$
   (b) $X = (A + B + \bar{C}) \cdot (A + \bar{B} + C) \cdot (A + B + C) \cdot (\bar{A} + \bar{B} + C)$
7. Simplify the combinatorial network of Fig. 2.51.

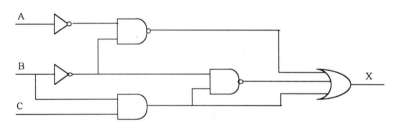

Fig. 2.51 Combinatorial network for Exercise 7.

8. Design a combinatorial network that takes a 4 bit binary input and yields an output only if the binary input is greater than 8.
9. Figure 2.32 showed the logic network to implement a half adder. Design the logic network for a device that adds two bits and provides a carry bit to the third position.
10. Figure 2.52 is a device that counts a binary coded decimal (BCD) input from a 4 bit register. Any combination of four BCD bits will light the appropriate lamp (light emitting diode). Assuming that only BCD inputs will occur, design the appropriate combinatorial network.

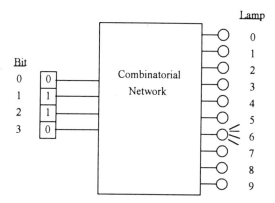

Fig. 2.52 BCD to lamp combinatorial network.

11. A seven-segment display is a device having seven parts labeled a to g as shown in Fig. 2.53. Each part is a light emitting diode (LED) that can be controlled by a combination of binary inputs. By lighting combinations of segments, the numbers 0 through 9 can be formed. The truth table for this device is in Fig. 2.53. Draw the combinatorial network.

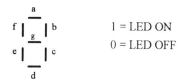

1 = LED ON
0 = LED OFF

INPUTS				OUTPUTS						
B_3	B_2	B_1	B_0	a	b	c	d	e	f	g
0	0	0	0	1	1	1	1	1	1	0
0	0	0	1	0	1	1	0	0	0	0
0	0	1	0	1	1	0	1	1	0	1
0	0	1	1	1	1	1	1	0	0	1
0	1	0	0	0	1	1	0	0	1	1
0	1	0	1	1	0	1	1	0	1	1
0	1	1	0	1	0	1	1	1	1	1
0	1	1	1	1	1	1	0	0	0	0
1	0	0	0	1	1	1	1	1	1	1
1	0	0	1	1	1	1	0	1	1

Fig. 2.53 BCD to 7 segment decoder.

12. A circuit widely used in data collection systems is called a multiplexer. The multiplexer in Fig. 2.54 allows the controller to sample any one of four inputs (I_1, I_2, I_3, I_4) of the multiplexer by selecting the appropriate input line using the control lines (C_1, C_2). Construct the combinatorial network for the multiplexer.

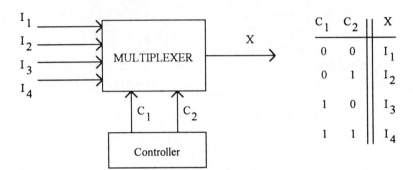

Fig. 2.54 Multiplexer.

13. Perform the following conversions:
    (a) 202_{10} to hexadecimal.
    (b) 10011_2 to hexadecimal.
    (c) AD_{16} to binary.
    (d) 56_{10} to binary.
    (e) 56_{16} to decimal.
14. Perform the following conversions:
    (a) 2's complement 101101 to decimal.
    (b) 47_{10} to 2's complement binary.
    (c) DAD_{16} to binary coded decimal.
    (d) 971_{10} to hexadecimal.
    (e) 101101_2 to hexadecimal.
15. The octal system uses the base 8. What is 422_8 in binary?
16. Perform the following arithmetic operations:
    (a) $AE + FE$.
    (b) $25_{10} - 41_{10}$ in 8-bit 2's complement binary arithmetic.
    (c) $ABCD_{16} + 138F_{16}$.
17. Perform the following arithmetic operations:
    (a) $1011_2 + 1101_2$
    (b) $1001_2 - 1010_2$
    (c) $AE - FE$ in 9-bit 2's complement arithmetic.
18. The position of a four-position index table is sensed using four switches: S1, S2, S3, S4. As the table rotates through the four quadrants, the switches are activated by pins on the index table. The chart below shows which

switches are activated in each of the four positions. Design a logic circuit that produces outputs indicating the quadrant the table is in based on the four switches as input.

Table Position (degrees)	S1	S2	S3	S4
0–90	0	0	1	1
90–180	0	1	1	0
180–270	1	1	0	0
270–360	1	0	0	1

19. For exercise 18, design a logic network to accomplish the same result using a minimum set of inputs.
20. Shown in Fig. 2.55 is a flip flop with two inputs. Write the truth table for this device.

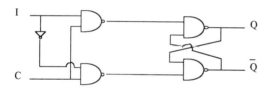

Fig. 2.55 Flip flop for Exercise 20.

21. Shown in Fig. 2.56 is a circuit known as an RS flip flop. One difficulty with the RS flip flop is that it has one undefined state. Develop the truth table for this device, indicating the undefined state.

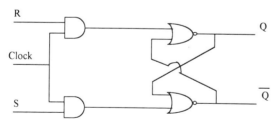

Fig. 2.56 RS flip flop.

22. An electronic device often used in control circuits is a comparator, shown below. A comparator has two inputs: a reference voltage and a voltage provided by an input device, such as a sensor. The logical behavior of the device can be described simply as follows:

if $V_{actual} < V_{ref}$, $X = 0$
if $V_{actual} \geq V_{ref}$, $X = 1$

V_{actual} ─────▷── X

V_{ref}

We wish to control a process in which temperature, humidity and pressure are being monitored using sensors that output actual voltages, V_{AT}, V_{AH} and V_{AP}, respectively. If actual voltages exceed maximum allowable (reference) voltages, an action will be taken. If $V_{AT} > V_{REFT}$ or $V_{AH} > V_{REFH}$, a fan will be turned on to lower the temperature of the process. If $V_{AT} > V_{REFT}$ and $V_{AH} > V_{REFH}$ and $V_{AP} > V_{REFP}$, the system will automatically shut down. A block diagram of the device is shown in Fig. 2.57. Draw the combinatorial network.

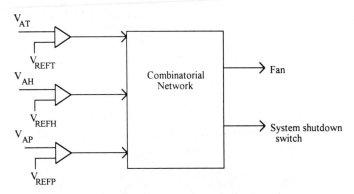

Fig. 2.57 System for Exercise 22.

23. A thermocouple monitoring a process is giving a voltage, V_{actual}, that is proportional to temperature. The process is controlled in the following manner. When V_{actual} exceeds a maximum allowable temperature, V_{refmax}, a fan is turned on. The fan stays on until the temperature goes down to a minimum acceptable temperature, V_{refmin}. An output signal, $S = 1$, turns on the fan and an output signal, $S = 0$, turns off the fan. The situation is shown in the block diagram in Fig. 2.58. Design an appropriate network. (Hint: the answer will require a sequential network).

EXERCISES

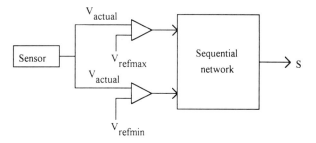

Fig. 2.58 Illustration for Exercise 23.

24. Consider the following program:

    LDAA #8A
    STAA 01F0
    LDAA #92
    ORAA 01F0
    STAA 01F0

   (a) After execution, what is the contents of address 01F0?
   (b) How long does it take to execute this program if the clock speed is 2×10^6 cycles per second?

25. A serial transmission is sending seven character bits with even parity and one stop bit as shown in Fig. 2.59. What ASCII character is being transmitted?

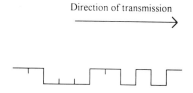

Fig. 2.59 Serial transmission of ASCII character.

26. Figure 2.60 shows a controller operating a conveyor and a reject diverter. Units on the conveyor pass under a vision system that can detect bad units. The system is controlled by a controller that is programmed using the 6800 assembly language. The machine vision system sends a bit to the controller each time an inspection is made. The controller's input port is a memory mapped bidirectional port at address $1003. Its data direction register is at address $1007. The following input signals apply:

Input	From	Message
PC0	Vision system	0 = nothing; 1 = board accepted
PC1	Vision system	0 = nothing; 1 = board rejected

The controller is responsible for the following control cycle:
(a) Turn the conveyor on.
(b) As long as there are no rejects, do nothing.
(c) If there is a reject, stop the conveyor and turn on the automatic reject device and delay for 100 microseconds. This will reject the part off the conveyor. Turn off the reject device and restart the conveyor.

The output port of the controller is located at address $1004 and is wired as folows:

Output	To	Message
PB0	Conveyor motor control	0: Motor off; 1: Motor on
PB1	Reject kickout device	0: Device off; 1: Device on

Write a program for the controller in 6800 assembly language.

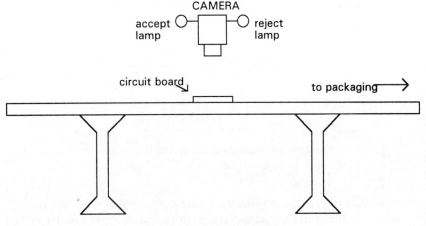

Fig. 2.60 In line inspection.

27. For the problem in exercise 26, add the following capability to the control program:

EXERCISES

(a) Maintain a running total of accepted boards, rejected boards and total boards inspected.

(b) Make a production run of 2000 boards, shutting the conveyor down afterwards.

28. For exercise 26, it has been decided that this controller should not be dedicated to the operation described. In addition, it should also service other processes. Since the controller is engaged in other services, it is decided to wire PC1 to the IRQ interrupt through an inverter. A bad unit will cause the sensor to interrupt the controller, which will then service the operations described in exercise 26. Assume that the interrupt routine is to be stored in address $0120 and that the interrupt vectors are stored in addresses $FFF2 and $FFF3. Show the programming required to accomplish this modification.

29. Shown in Fig. 2.61 is a filling line consisting of a conveyor, a filling station and a weighing station. Packages move through the sequence of activities one at a time. At each station there is a proximity sensor to indicate the presence of a package to be filled or to be weighed. When a package is present at the input station, the controller turns the conveyor on and moves the package to station 1, where the filling machine deposits material into the package. When this is complete, the conveyor indexes the package to the next station, where it is weighed. An output bit from the checkweigher indicates whether the package has enough material in it or not. If the package is underweight, it is diverted by the controller to a rework area. If the package has the minimum required weight, it proceeds directly to the sealing operation. It is assumed that packages go through the system completely one at a time.

It is your assignment to write the control software for this application. The controller is responsible for starting and stopping the conveyor as well as initiating the filling operation, weighing operation and diverter gate that diverts underweight trays to the rework area. The following are the input and output bit assignments of the controller:

Input	Description	Message	
I0	start conveyor push button	0: nothing	1: start conveyor
I1	proximity sensor at filling station	0: no package	1: package present
I2	proximity sensor at weigh station	0: no package	1: package present
I3	filling complete signal from fill station	0: not complete	1: filling complete
I4	weighing complete signal	0: not complete	1: weigh complete
I5	prox sensor reject lane		1: part passing
I6	prox sensor on accept lane		1: part passing
I7	result of weighing operation	0: acceptable	1: underweight

Output	Description	Message	
O1	conveyor motor control	0: stop conveyor	1: start conveyor
O2	fill package	0: no action	1: fill package
O3	weigh package	0: no action	1: weigh package
O4	diverter control	0: diverter off	1: diverter on

The filler and conveyor have their own controllers that control their operating cycles when the supervisory controller requests them to execute their cycles. The above operation requires what is called a "full handshake". For example, when the controller requests that a package be filled by putting a "1" on O2, the controller must wait for a return signal from the filler indicating that the package has been filled (I3 = 1). At that time the controller disables the fill package signal. When all requested operations are complete, the conveyor is restarted and packages are moved foward. The input port is at address $1003 and the output port is at address $1004.

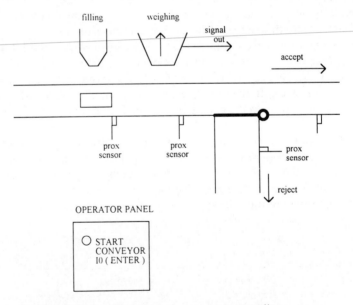

Fig. 2.61 Filling operation on a conveyor line.

FURTHER READING

Campbell, J. (1984) *The RS-232 Solution*, SYBEX, Inc., Berkeley.
Cline, B. E. (1983) *An Introduction to Automated Data Acquisition*, Petrocelli Books, Inc., New York.

FURTHER READING

Evans, A. J., Mullen, J. D. and Smith, D. H. (1985) *Basic Electronics Technology*, Texas Instruments Information Publishing Center, Fort Worth, Texas.

Gibson, G. A. and Liu, Y. (1980) *Microcomputers for Engineers and Scientists*, Prentice-Hall, Inc., Englewood, Cliffs, New Jersey.

Greenfield, J. D. (1991) *The 68HC11 Microcontroller*, Saunders College Publishing, Harcourt Brace, Philadelphia, Pennsylvania.

Hall, D. V. (1992) *Microprocessors and Interfacing*, 2nd edn, Macmillan/McGraw-Hill, Glencoe, Illinois.

Hilburn, J. L. and Julich, P. M. (1976) *Microcomputers/Microprocessors: Hardware, Software, and Applications*, Prentice-Hall, Inc., Englewood Cliffs, New Jersey.

Hintz, K. J. and Tabak, D. (1992) *Microcontrollers*, McGraw-Hill, Inc., New York, New York.

Lipovski, G. J. (1988) *Single and Multiple-Chip Microcomputer Interfacing*, Prentice-Hall, Inc., Englewood Cliffs, New Jersey.

Motorola (1991) *M68HC11 Reference Manual*, Motorola, Inc.

Schmitt, N. M. and Farwell, R. F. (1983) *Understanding Electronic Control of Automation Systems*, Texas Instruments, Inc., Dallas, Texas.

Tocci, R. J. and Laskowski, L. P. (1987) *Microprocessors and Microcomputers: Hardware and Software*, Prentice-Hall, Inc., Englewood Cliffs, New Jersey.

3 Sensors and automatic data acquisition

3.1 INTRODUCTION

In Chapter 2 we focused on the architecture of the brains of a control system, the microprocessor and associated memory and input/output devices. If the computer is the brains, sensors are the eyes and ears of the control system. In computer control applications in manufacturing, sensors are as important as the computer itself. Sensors can be used to provide real time information for directly controlling processes as well as provide information for data logging purposes; for example, to provide a count of the daily units produced off a particular manufacturing line.

The topic of sensor technology is vast. Indeed, a large number of sensors for manufacturing, laboratory testing, process control and other areas are commercially available. In this chapter we will focus on the most commonly used sensors found in manufacturing applications.

Figure 3.1 is a classification framework for some of the most common sensors used in manufacturing. They can be broadly classified in two categories: **discrete event** and **continuous**. A discrete event, or on/off sensor, changes its state based on the occurrence of some external event. The event may be the passing of a part by a point along a conveyor or the movement of a mechanical device beyond some prespecified limit. These sensors typically only give knowledge of two states based on the condition being sensed. In the first example, the part is either present or not present. In the second case the device is either within the spacial limit or it is not. Discrete event sensors used in manufacturing are typically based on either mechanical, electrical or optical technology. Some of these sensors will be described in section 3.2.

A continuous sensor measures the magnitude of an attribute of interest of the physical process that is being monitored. The process may be the temperature of a heat treating process, the pH in a chemical process, or the speed of a conveyor line. These sensors provide information over the continuous range of operation of the process and they are commonly used in continuous control applications, where the process is being regulated based on continuously sensed attribute data. Continuous sensors used in manufacturing are commonly based on electrical, optical and acoustic technologies. Several of these sensors will be described in a later section.

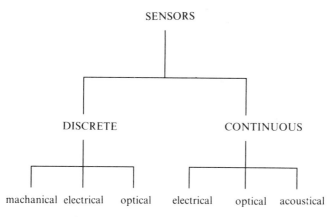

Fig. 3.1 A classification of sensor types.

3.2 DISCRETE EVENT SENSORS

In this section we shall describe some of the more common discrete event sensors used in manufacturing. These include mechanical limit switches, proximity sensors and photoelectric sensors.

3.2.1 Mechanical limit switches

Mechanical limit switches typically consist of a mounted actuator arm that operates a set of electrical contacts when the arm is displaced. Two examples are shown in Fig. 3.2. Figure 3.2a illustrates the operation of a **lever-type** limit switch and Fig. 3.2b illustrates a plunger, or **push-type**, limit switch. In the case of the lever type, the actuator arm is a rod (1) connected to a lever shaft (2), which is free to rotate when the rod is displaced. When the forces displacing the rod are removed, the lever shaft is returned to its normal position by a return spring (3). The lever shaft has a roller mounted on its bottom (4), which rotates a rocker (5) as it changes position from right to left. This mechanical action operates one or more sets of contacts, which are mounted on the other side of the limit switch, as shown in the back view. The rocker shaft is connected through the housing to the contact lever assembly (6), the head of which moves a set of electrical contacts (7).

The electrical contact may either be closed or open initially. The action of the actuator and lever arms takes it from its normal, or deactivated, state to the other state. Hence, a normally open limit switch will be closed when activated and a normally closed limit switch will be open when activated. Figure 3.2b illustrates the more direct action of a push-type limit switch. It shows a set of contacts operated from the depression of the contact lever assembly. Contact

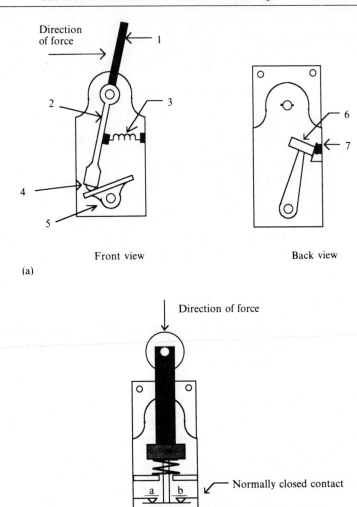

Fig. 3.2 Mechanical limit switches. (a) Lever type; (b) push type.

set a–b is normally closed; contact set c–d is normally open. When the lever is depressed, each contact goes to its opposite state. When installed, the user wires the appropriate contact pair back to the controller, which distinguishes the state of the system by reading the voltage or current (on/off) supplied through the contact.

Limit switches come in several varieties and designs; Fig. 3.2 simply shows two concepts. They are designed for heavy duty applications in which there is physical contact between the actuator and the process being sensed. For example, limit switches are often used on machine tools to limit the travel of a machine axis. They are sometimes used in materials handling applications e.g. to indicate the passage of a part along a conveyor. They are typically designed to handle relatively high voltages, both AC and DC. This means that they cannot be directly wired to the input port of a computer without having their signal converted to TTL levels.

3.2.2 Proximity switches

The term proximity switch (sometimes called proximity sensor) refers to a non-contact sensor that works on the principle of inducing changes in an electromagnetic field. The proximity switches most commonly used in the manufacturing environment are the **inductive proximity switch** and the **capacitive proximity switch**.

The principle of an inductive proximity switch is shown in Fig. 3.3. The proximity switch uses an oscillator circuit to generate a small, high frequency radio signal emitted from the end face of the sensing head. When a metallic object enters the range of the sensor, small circular currents, called eddy currents, are induced on the surface of the metal. This causes a change on the load of the oscillator, which is recognized by the control circuit of the proximity switch. The control circuit sends an output signal to a controller, indicating the presence of an object.

One of the shortcomings of the inductive proximity switch is that it can only sense metal objects. The capacitive proximity switch, on the other hand, can sense non-metallic objects as well. It uses a resistor/capacitor (RC) oscillator to generate a directed magnetic field. Introducing an object within the magnetic field causes a change in capacitance, which is detected by the control circuitry, which in turn operates an electronic switch that outputs a signal to the controller.

Proximity switches have relatively short sensing ranges, typically from 1 to 60 mm; therefore, they must be used in situations where the target is allowed to

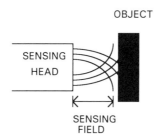

Fig. 3.3 Proximity switch.

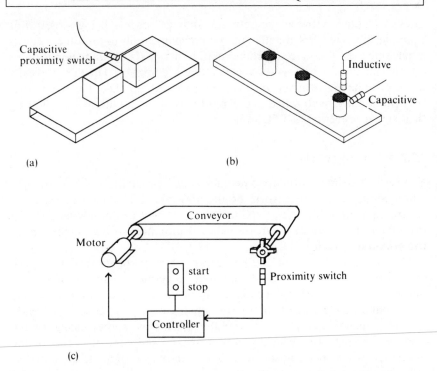

Fig. 3.4 Applications of the proximity switch. (a) Capacitive proximity switch counting cartons; (b) inductive proximity switch detecting presence of metal cap; (c) proximity switch monitoring operation of conveyor.

come close to the sensor. Figure 3.4 shows some typical applications for proximity switches. In Fig. 3.4a, a capacitive proximity switch is detecting cardboard cartons moving along a conveyor. In Fig. 3.4b a proximity switch is detecting the presence or absence of a metallic cap on passing jars of food product. This is a good application for an inductive proximity switch since a jar without a cap will not be sensed, as it would be by a capacitive proximity switch. Note the requirement for a capacitive proximity switch to sense the presence of the jar. Without this sensor, the controller would not know when a jar without a cap has passed. In Fig. 3.4c, a proximity switch is detecting the motion of a metallic toothed wheel on a conveyor shaft. Each time a tooth passes through the sensing range of the proximity switch, motion is detected. Such a signal can be used to inform the motor controller whether or not a jam has occurred that stopped the conveyor. If the motor drive is on and there is no intermittent signal coming back from the switch, it indicates that a jam has occurred and power to the motor drive should be shut down before the motor burns out.

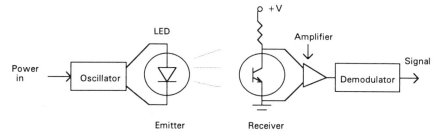

Fig. 3.5 Principle of a photoelectric sensor.

3.2.3 Photoelectric sensors

Photoelectric sensors are non-contact devices that output a signal in response to the interruption of a light beam. The components of a photoelectric sensing system are shown in Fig. 3.5. The two main components are the emitter and the receiver. The light source is a light emitting diode (LED). An LED is a solid state semiconductor that emits light when current flows through it. LEDs are manufactured to produce light in the visible range and in the near infra-red range. The light source is paired with a receiver, which is a light sensitive transistor, called a phototransistor. The reader will recall from Chapter 2 that a transistor conducts when its base is forward biased. This is done by applying a small amount of current on the base lead. A phototransistor operates in the same fashion except that the base is biased by the energy from a light source incident on it. Phototransistors are manufactured to be sensitive to light within the spectrum of the emitter.

A photoelectric sensor system comes with an oscillator that modulates, or pulses, the LED on and off at very high frequencies. The receiver is tuned to the same frequency, which allows it to differentiate between light from the emitter and ambient light. This is analogous to frequency modulation in FM radio, which enables a listener to tune precisely to a given station.

Photoelectric sensors are designed for use in four basic sensing modes: opposed, retroreflective, diffuse, and convergent. The appropriate sensing mode depends on the application.

Figure 3.6a shows the **opposed sensing mode**, sometimes referred to as the transmitted beam mode. In this mode, the emitter and receiver are positioned opposite one another. When an object passes through the beam, interrupting it, the base of the phototransistor receiver is de-energized. In the opposed sensing mode the object must be large enough to cover the entire beam width. Otherwise, some light will always be incident on the receiver and the object will not be detected. For that reason, manufacturers of photoelectric sensors provide a control on the emitter to allow a user to aperture the light down to a fairly narrow beam. Although the opposed sensing mode is considered very reliable, it requires the expense of purchasing separately housed emitter/receiver pairs with additional wiring considerations. This is avoided in the retroreflective mode.

Figure 3.6b illustrates the **retroreflective sensing mode**. In this mode the emitter and receiver are housed in the same package. The light beam is established by bouncing it off a special target, which returns it to the receiver. Again, the principle is that an interruption of the light beam indicates the presence of an object.

Although retroreflective sensors are more economical, there are some additional considerations. Objects that have a highly reflective surface may return the light beam and pass undetected. Retroreflective sensors provide much less light intensity and their effectiveness can deteriorate when used in an environment where dirt collects on the reflector. On the other hand, when the target is large and non-reflective and the environment is relatively dirt free, it offers a good solution.

Figure 3.6c illustrates the **diffused sensing mode**. This mode has the emitter and receiver housed in the same package. In diffused sensing, light is emitted at

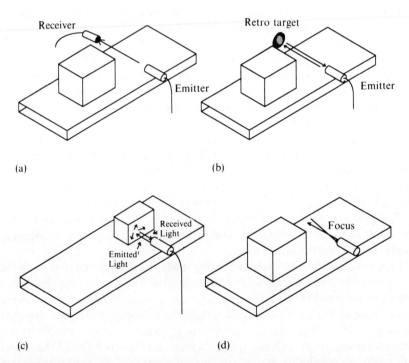

Fig. 3.6 Illustrated photoelectric sensing modes along a conveyor. (a) Opposed sending mode; (b) retro-reflective sensing mode; (c) diffuse sensing mode; (d) Convergent sensing mode.

some wide angle and there is no reflective target. Instead, the return light comes from the object itself, which must have a reflective surface. The receiver will pick up some portion of the returned light, which biases the phototransistor base. The logic is opposite to the opposed and retroreflective sensing modes; the presence of an object makes the beam rather than breaks the beam.

Convergent beam sensing is illustrated in Fig. 3.6d. In this mode, the emitter and receiver are housed in the same package. The emitter light is focused at a specific point, which limits the sensing depth of field. An object passing through that range will reflect light back to the receiver, which is tuned to a specific light requirement based on the reflectivity of the object. The importance of the convergent sensing mode is that a user can focus detection on a very specific area. For example, if it is important to check an assembly moving along a conveyor for the presence of a part in a specific place, the convergent sensing mode is a good choice.

Photoelectric sensors have the capability of being used over a fairly long distance. For example, distances over 100 feet are possible in an opposed sensing mode.

3.2.4 Sensor arrays

Another configuration for using photoelectric sensors is the sensor array, shown in Fig. 3.7. The sensor array provides several binary inputs, which are scanned by the controller to determine which are on and which are off. As objects traverse the field of view, the pattern of obstructed light can be used to identify the size or orientation of an object. A typical use is to sort objects by size on a conveyor, as shown in Fig. 3.7.

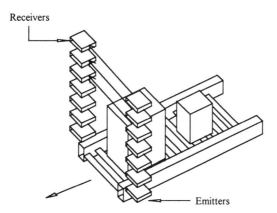

Fig. 3.7 An application of sensor arrays.

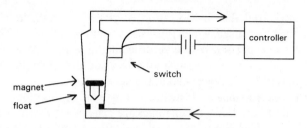

Fig. 3.8 Implementation of a flow switch.

3.2.5 Fluid flow switches

A discrete event sensor commonly used in process industries is the fluid flow switch. This device is analogous to a limit switch in mechanical systems. It is usually employed as a device to detect when a fluid travelling in a pipe is over a specified volumetric flow rate. Such a sensor can be used to govern the speed of an upstream pump, reducing it when the flow is too fast.

There are different implementations of a fluid flow switch. One implementation is shown in Fig. 3.8. Here a float is placed in a tapered column that is in line with the fluid flow. A magnetic ring is seated on the float and, in the absence of pressure, the movable float is retained in the seated position. Water pressure raises the float proportional to the flow rate. A switch is positioned at a height where the float will be displaced when the triggering flow rate is reached. Typically the switch is a reed switch, which consists of two leaf springs sealed in glass. These leaf springs come together when subjected to a magnetic field. This closes the circuit, providing an input signal to the controller.

3.3 CONTINUOUS SENSORS

In Chapter 1 we introduced the concept of closed loop continuous control as illustrated in Fig. 1.3. This has been reproduced, with some additions, in Fig. 3.9. In the following sections we focus on those components of the system indicated in the dotted enclosure. Sensors used in closed loop continuous control are the instruments that feed back process parameter measurements to the controller. Typically, these are analog devices that measure the magnitude of some parameter of the system output, such as speed, position or temperature. The controller, a digital device, requires that the information provided by the sensor be converted to digital form before it can be acquired and recorded by the processor of the controller. The device that performs this conversion is an **analog to digital (A/D) converter**.

In a similar manner, actuators are usually driven by voltage or current levels of a magnitude higher than that which can be delivered by a computer. If the

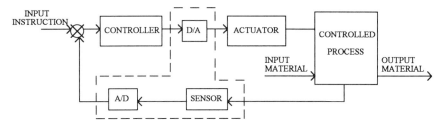

Fig. 3.9 Closed loop control with analog to digital and digital to analog conversion.

computer is to control these voltage levels, there must be a mapping between a digital output code of the controller and an analog voltage or current level. The device that performs this conversion is a **digital to analog (D/A) converter**. Although the D/A converter is used for actuator control, not sensing systems, it is convenient to discuss it at this time because the concepts involved are closely related to A/D converters.

The next section provides an overview of a continuous sensor system. This is followed by a description of A/D and D/A signal conversion and a discussion of some important sensor technologies.

3.3.1 An overview of a sensor system

The term **continuous sensor** is used to describe a device that converts one measured physical quantity into another that is proportional to the measured physical quantity. The measured physical quantity might be position, velocity or temperature; the converted proportional physical quantity is typically one that can be used in an electronic circuit, such as electrical resistance.

Conceptually, the components of a sensing system and their relationship to a digital controller are shown in Fig. 3.10. Each component of the chain has a unique purpose, which shall be briefly explained.

The sensor takes the actual physical input, such as force or temperature, and provides an output that can be used by an electronic measurement circuit. The measurement circuit is used for calibration and reading of the sensor output. The output of the measurement circuit will be a voltage or current that is proportional to the physical property of the system being measured by the sensor. For example, it may be a voltage whose magnitude is proportional to a force being measured at the sensor/process interface.

Sometimes a signal will have electronic noise or be too weak to be processed directly from the measurement circuit. In such cases the signal may be filtered to remove the noise or may be amplified for further processing. This is the role of signal conditioning.

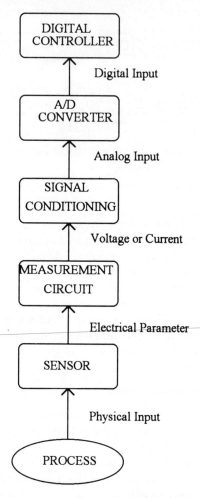

Fig. 3.10 Components of a sensing system.

Since computers and controller processors are digital devices, any analog signal will have to be digitized before it can be read by the computer. An A/D converter provides this conversion. In cases where the sensor signal is already in digital form, such as that emitted from an optical encoder, this step is unnecessary. Finally, the signal is presented to the computer for processing.

In the next sections we shall describe some of the important steps of this process in detail. For the purposes of illustration, we begin by describing the process in terms of a specific sensor, the potentiometer, which is often used as a position sensor.

3.3.2 Some basic principles of instrumentation: example of a potentiometer

An often-used position sensor is the potentiometer. The potentiometer is composed of a resistor and a contact slider that allows position to be made proportional to resistance. There are linear and rotary potentiometers depending on whether the displacement to be measured is linear or angular.

Figure 3.11 is a schematic of a linear potentiometer. Figure 3.11a illustrates the relationship between change in position (ΔX) and change in resistance (ΔR) between points a and b. If a voltage (V_i) is applied across the entire resistor, as shown in Fig. 3.11b, the voltage drop between terminals a and

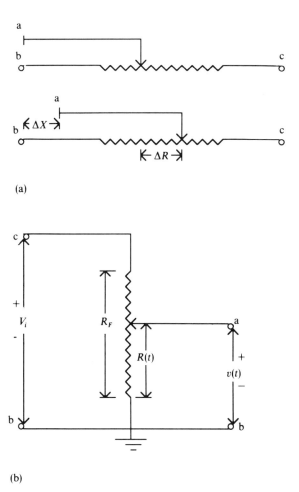

Fig. 3.11 Principle of a linear potentiometer. (a) Position displacement and resistance change; (b) voltage division across potentiometer.

b will be proportional to the resistance between a and b. Hence, with proper calibration, a linear relationship can be set up between a physical phenomenon (displacement or position) and an electrical measure (resistance or voltage).

Figure 3.11b is a simple sensor that converts a physical measure into an electrical parameter as described in Fig. 3.10. The output voltage at time t, $v(t)$, can be used as the input for measurement and further processing. The voltage $v(t)$ can be written:

$$v(t) = \frac{R(t)}{R_F} V_i$$

or

$$\frac{v(t)}{V_i} = \frac{R(t)}{R_F}, \qquad (3.1)$$

where:
$v(t)$ = the voltage reading at time t
$R(t)$ = the resistance between wiper and ground at time t
R_F = the full scale resistance
V_i = the input voltage.

Since

$$\frac{v(t)}{V_i} = \frac{R(t)}{R_F},$$

the voltage reading is a simple linear function of the slider position.

3.3.3 Avoiding measurement nonlinearity

Unfortunately, to measure $v(t)$ requires a meter or further circuitry to pass the signal up the chain of Fig. 3.10. This circuit will present a load across terminals a–b which can create nonlinearities in the measurement. Consider the simple circuit shown in Fig. 3.12. The value R_L is the impedance of the load from the measurement circuit itself. The value $v(t)$ is now given by:

$$v(t) = V_i \frac{R_{eq}}{(R_F - R(t)) + R_{eq}} \qquad (3.2)$$

where:

$$R_{eq} = \frac{1}{1/R(t) + 1/R_L} = \frac{R(t)R_L}{R(t) + R_L}$$

$$\therefore \frac{v(t)}{V_i} = \frac{R(t)R_L}{[R(t) + R_L][R_F - R(t)] + R(t)R_L}$$

$$\frac{v(t)}{V_i} = \frac{R(t)R_L}{R(t)R_F - R(t)^2 + R_L R_F}$$

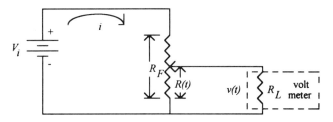

Fig. 3.12 Potentiometer with voltmeter measuring output.

Multiplying numerator and denominator through by $1/(R_L R_F)$,

$$\frac{v(t)}{V_i} = \frac{R(t)/R_F}{[R(t)/R_L] - [R(t)^2/R_L R_F] + 1} \quad (3.3)$$

There are two things to note about equation 3.3. The first is that there is a nonlinear output/input relationship. Nonlinearity is undesirable in a measurement system. Any calibration that assume linearity between the wiper displacement and the output voltage will be in error. If equation 3.3 is the correct output/input relationship and the calibration is based on equation 3.1, the magnitude of the fractional error can be computed:

$$\text{Fractional error} = \frac{\text{Eq (3.3)} - \text{Eq (3.1)}}{\text{Eq (3.3)}}$$

The second point to note is that as

$$\frac{R(t)}{R_L} \to 0, \quad \frac{v(t)}{V_i} \to \frac{R(t)}{R_F}.$$

Hence, if the measurement circuit is designed so that the load impedance is very high, the linearity of the measurement can be improved. For this reason, electronic measurement instruments, such as voltmeters, are designed with a high impedance relative to the circuit to be measured. In Fig. 3.10, the role of the measurement circuit is to ensure the best linear relationship between the sensor electrical output and the subsequent signal processing.

When analog measurement systems are purchased from vendors, they are usually complete instrumentation packages. They include the sensing element combined with a measurement circuit that minimizes nonlinearity. Sometimes the instrumentation will include signal conditioning that provides a clean electrical output. The term **transducer** is often used to describe the entire measurement system that converts the measured input to a usable electronic signal.

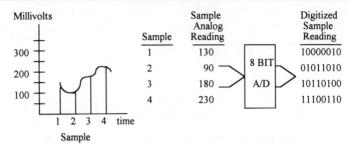

Fig. 3.13 Analog to digital conversion process.

3.3.4 Analog to digital signal processing

The electrical output of a potentiometer transducer is typical of many transducers: it is a continuous, or analog, signal. The signal must be digitized before it can be read by a computer. An analog to digital (A/D) converter is a device that transforms a continuous analog signal into digital form. The steps by which this is done are as follows:

1. The continuous signal is periodically sampled to form a series of discrete time samples.
2. Each discrete analog sample falls within one of a finite number of predefined amplitude levels called **quantizing levels**, which consist of discrete voltages over the range of the device.
3. The amplitude levels are encoded into digital form and presented for acquisition by the digital processor of the controller.

Figure 3.13 is a generalized illustration of the process. In this figure, the analog signal from the transducer is sampled by the A/D converter along the time axis of the analog signal. The A/D converter converts these samples into binary code. Figure 3.13 illustrates an 8-bit binary output, which means that this A/D converter is capable of classifying analog voltages into 256 discrete levels. It is this digitized reading that is the input to the processor memory of the controller. We shall discuss the process of converting signals between analog and digital systems in more detail in section 3.4.

3.4 INTERFACING A DIGITAL CONTROLLER WITH AN ANALOG WORLD

One of the features common in an industrial controller is its ability to interface to analog devices using A/D and D/A converters, sometimes designed into the controllers, themselves. The block diagram of Fig. 3.14 illustrates the interface organization. On the input side, a measurement process provides an electronic

INTERFACING DIGITAL CONTROLLER WITH ANALOG WORLD

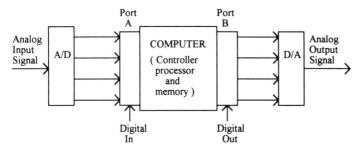

Fig. 3.14 Controller interfacing to analog devices.

signal in analog form. This signal is digitized into a binary string, which is input to the digital processor and memory of the controller. In Fig. 3.14 we illustrate a parallel input. Signals that drive actuators on the output side are controlled by binary digital outputs. These are converted to analog signals using a D/A converter. In discussing these conversion devices, we start with the D/A converter.

3.4.1 Digital to analog (D/A) converters

The basic process of taking a binary number and converting it to a voltage level is illustrated for a 4-bit binary D/A converter in Fig. 3.15a. With 4 bits provided by the computer, $2^4 = 16$ possible input states can be presented to the D/A converter. The desired range of the output voltage must be designed into the D/A device. For Fig. 3.15, the full range of the device is 0 to 11.25 volts. The input binary count can range between 0–15. Since there are 15 increments over the full range of the 11.25 volts, each increment of the binary count equals $11.25/15 = 0.75$ volts.

Figure 3.15b illustrates the relationship between the binary input and analog output. Each increment of the binary count adds 0.75 volts to the output voltage. The term **step size** is used to refer to the smallest increment of output (voltage or current) that can be controlled by the binary input. The step size for Fig. 3.15 is 0.75 volts. Step size is related to **percent resolution**, which is the percent of full range voltage or current occupied by one step. Percent resolution can be computed from either the step size or the number of bits (n) of input used over the full range as follows:

$$\% \text{ resolution} = \frac{\text{step size}}{\text{full range}} \times 100\% \qquad (3.4)$$

$$\% \text{ resolution} = \frac{1}{2^n - 1} \times 100\% \qquad (3.5)$$

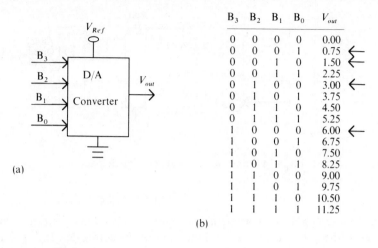

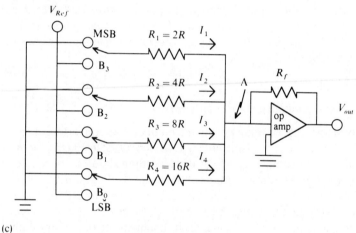

Fig. 3.15 Digital to analog conversion process. (a) 4-bit D/A block diagram; (b) binary input and analog output; (c) implementation of a 4-bit D/A converter.

EXAMPLE 3.1

A 6-bit D/A converter gives an output voltage of 8.625 volts for an input of 010111. What is the step size, the full range voltage, and the percent resolution?

Answer

The value 23_{10} yields the output 8.625 volts. Therefore, the step size is:

$$\frac{8.625 \text{ volts}}{23} = 0.375 \text{ volts/step}$$

INTERFACING DIGITAL CONTROLLER WITH ANALOG WORLD

For a six-bit input, the largest attainable value is 63_{10}. Therefore, the full range voltage is:

$$63 \times 0.375 = 23.625 \text{ volts}$$

Therefore, the percent resolution is:

$$\% \text{ resolution} = \frac{\text{step size}}{\text{full range}} \times 100 = \frac{0.375}{23.625} \times 100 = 1.587\%$$

$$\% \text{ resolution} = \frac{1}{2^n - 1} \times 100 = \frac{1}{2^6 - 1} \times 100 = 1.587\%$$

A re-examination of Fig. 3.15b will show an interesting point. There is a relationship between the binary position weights and the output voltages. This is indicated by the arrows on the side of Fig. 3.15b. In particular, a '1' in each successive binary position results in a doubling of the output voltage. This is a clue to the way in which D/A converters are constructed, which is illustrated in Fig. 3.15c.

Each binary output signal from the computer at five volts controls an electronic switch. Each electronic switch is used to connect or disconnect a branch to a reference voltage, V_{Ref}. Current will flow in connected branches. The op amp is a device that produces a weighted sum of the input voltages.

The op amp voltage, V_{out}, is the output voltage of a ladder of resistors which are electronically switched (connected) to a reference voltage, V_{Ref}. Hence, by placing the appropriate binary code on $B_0 - B_3$, the voltage level V_{out} can be produced.

The values of the resistors on the ladder are incremented in binary powers, i.e. $2R, 4R, \ldots, 2^n R$. Hence, from the law of passive linear circuits, the current in each branch of the ladder is:

$$I_1 = V_{Ref}/R_1 = V_{Ref}/2R$$
$$I_2 = V_{Ref}/R_2 = V_{Ref}/4R$$
$$I_3 = V_{Ref}/R_3 = V_{Ref}/8R$$
$$\vdots$$
$$I_n = V_{Ref}/R_n = V_{Ref}/2^n R$$

The current at point A in the circuit is the summation of the current through each branch:

$$I_A = (V_{Ref}/R) \sum_{i=1}^{n} a_i 2^{-i}$$

where:

I_A = the current at the junction point A
V_{Ref} = the reference voltage

R = the resistance basis
a_i = a binary indicator (0 or 1) of whether the ith bit position is off or on, where the first position is the most significant bit (MSB)
n = the number of bits in the output register and, therefore, the number of branches in the ladder.

Finally, $V_{out} = I_A R_f$.

The above equations can be combined to yield the following function for a D/A converter:

$$V_{out} = V_{Ref} \frac{R_f}{R} N \qquad (3.6)$$

where $N = (a_1/2^1) + (a_2/2^2) + \cdots + a_n/2^n$.

The appropriate D/A conversion is obtained by choosing V_{Ref}, R_f, and R. When $R = R_f$,

$$V_{out} = V_{Ref} \sum_{i=1}^{n} a_i 2^{-i}$$

EXAMPLE 3.2

In Fig. 3.15, $R = R_f = 1$ k ohms. What is the appropriate value of V_{Ref}?

Answer

When $N = 15/16$, $V_{out} = 11.25$ volts. Therefore,

$$V_{out} = V_{Ref} \frac{R_f}{R} N,$$

$$V_{Ref} = 11.25 \frac{16}{15} = 12 \text{ volts}$$

The output voltage range can be changed by appropriate changes in the parameters of equation 3.6. For example, the range of V_{out} can be doubled by increasing V_{Ref} or R_f by twice.

EXAMPLE 3.3

In Fig. 3.15, $R_f = R$. What is the value of V_{out} when $B_0 = B_2 = 1$ and $B_1 = B_3 = 0$?

Answer

$$V_{out} = V_{Ref} \sum_{i=1}^{n} a_i 2^{-i}$$

$$V_{out} = 12[(0)2^{-1} + (1)2^{-2} + (0)2^{-3} + (1)2^{-4}]$$

$$= 12[0.3125]$$

$$= 3.75 \text{ volts}$$

3.4.2 Analog to digital conversion

When a sensor provides an analog voltage input to the controller, it will be necessary to convert the signal to digital form for computer processing. This is the role of the A/D converter. This is accomplished by sampling the analog input at discrete intervals of time and mapping each sample into one of the discrete quantizing levels of the converter. An A/D converter has 2^n discrete quantizing levels, where n is the number of bits in the register of the A/D converter.

Figure 3.16a illustrates a 4-bit A/D converter, with a full range voltage of 11.25 volts, i.e. it is designed to be used over the range 0 to 11.25 volts. The mapping of voltage into a binary count is shown in Fig. 3.16b. Unlike a D/A converter that maps a binary count to a specific voltage, an A/D converter maps a voltage range to a binary count. In general, the analog signal will fall into one of the quantizing levels of the converter. There are 16 quantizing levels and 15 incremental changes of 0.75 volts each.

The resolution of an A/D converter is determined by the step size, which is 0.75 volts in Fig. 3.16. As in the case of D/A converters, the percent resolution can be computed using equations 3.4 and 3.5. Therefore, the percent resolution is $(0.75/11.25) \times 100 = 6.67\%$. The term **accuracy** is used to describe the worst case error between the actual analog input signal and the recorded value as determined from the digital reading. In general, the accuracy is one half the resolution. This is illustrated in the conversion process of Fig. 3.16c. During the A/D conversion process, the converter samples the input signal, as shown on the left of Fig. 3.16c. The values of the continuous signal that is being sampled are in column 1 of the table. The actual data point that is recorded by the controller depends on the resolution of the A/D converter. Column 2 of Fig. 3.16c shows the binary count recorded by the computer based on conversion to the appropriate quantizing level. Column 3 shows the data point that the binary count is assumed to represent. The data point is the midpoint of the quantizing level. Since the recorded data point is the midpoint of the quantizing level, the worst case error is ± 0.375 volts, which is the accuracy of this particular A/D converter.

SENSORS AND AUTOMATIC DATA ACQUISITION

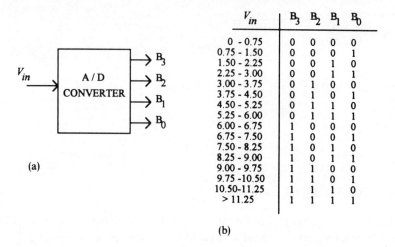

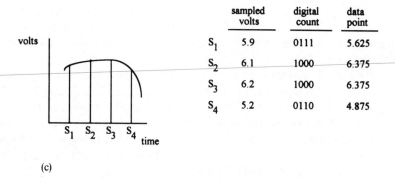

Fig. 3.16 Analog to digital conversion process. (a) 4-bit A/D block diagram; (b) analog input and binary output; (c) sampling process and stored data.

EXAMPLE 3.4

A typical A/D converter has 12-bit resolution and a full range of 10 volts. What is the percent resolution and voltage resolution of this device? What stored data point will I get for a sampled value of 7.3 volts?

Answer

$$\% \text{ resolution} = \frac{1}{2^n - 1} \times 100\% = \frac{1}{2^{12} - 1} \times 100\% = 0.02442\%$$

$$\text{voltage resolution} = 10^v \times 0.0002442 = 0.002442^v$$

INTERFACING DIGITAL CONTROLLER WITH ANALOG WORLD

For a reading of 7.3 volts, we can compute the quantizing level as:

$$\frac{7.3^v}{0.002442} = 2989.353$$

The reading will round to the 2989th quantizing level. Therefore, the stored reading will be:

$$(2989)(0.002442) + (1/2)(0.002442) = 7.300359 \text{ volts.}$$

The 2989th quantizing level has a range of 7.299138 to 7.30158. The midpoint is 7.300359.

There are different ways of implementing an A/D converter in hardware. One of the most widely used methods is based on taking 'successive approximations' of the analog signal.

Figure 3.17a shows the components of a simple A/D converter that uses the **successive approximation method**. The key device in the circuit is a comparator, which compares the actual analog signal at V_2 (unknown) with a known signal for comparison, fed in at V_1. Figure 3.17b shows the transfer function of the comparator. The output voltage is either positive $V_2 \geqslant V_1$ or negative $V_2 < V_1$. This results in a binary (1,0) input to the A/D register. The A/D data is generated by placing a known binary number in the A/D register, thus creating a known voltage once this binary count is converted by the D/A converter. The comparator tells the control logic whether the unknown voltage is above or below the known voltage. An organized search procedure is used to bracket the unknown voltage.

Figure 3.18 illustrates the process of approximating the unknown voltage using a four-bit A/D register. The A/D converter has 16 quantizing levels and, for illustrative purposes, we assume the voltage range is 0–11.25 volts and the actual input signal is 6.5 volts. Each quantizing level has a span of 0.75 volts. The approximation process sets each bit of the A/D register sequentially, starting with the most significant bit. Sample 1 sets bit 3, which is a comparator voltage of $8/15 \times 11.25$ volts $= 6.0$ volts. Since $V_2 > V_1$, bit 3 is held at a logic level 1. With bit 3 set, bit 2 is set $= 1$ for the next trial. This yields a comparator voltage of $12/15 \times 11.25$ volts $= 9.0$ volts. Since $V_2 < V_1$, bit 2 is cleared. The process continues until all bits of the register are loaded. The digital count for the unknown analog voltage is found by reading the A/D register. It is this digital count that is read and stored by the computer. The digital count can be interpreted as a voltage and recorded by the appropriate computation within the computer.

3.4.3 Other sources of measurement error

We have described accuracy as a measure of the degree to which the stored data point and the actual analog signal are close to one another based on resolution, which depends on the number of bits in the converter. There are

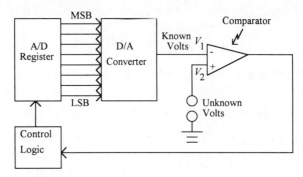

(a) Block Diagram of A/D Converter

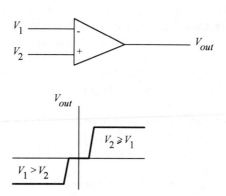

(b) Transfer Function of a Comparator

Fig. 3.17 Hardware organization of A/D converter. (a) Block diagram of A/D converter; (b) transfer function of a comparator.

other measures of the accuracy of the measurement which manufacturers of D/A and A/D converters usually specify.

Due to variations in the components used to build a converter, there will be some uncontrollable variations in the voltage output for a binary input among D/A units produced. The manufacturer describes the worst case variation from an 'ideal' converter as the accuracy of the model. So, for example, an accuracy of $\pm 0.05\%$ means that a given value may vary from its ideal value by $\pm 0.05\%$. This variation is in addition to the accuracy due to the number of bits.

Linearity error describes how much the step size of a converter deviates from its ideal step size. Thus, if a converter has a step size of 0.75 volts and a linearity error of $\pm 0.05\%$, the step size could be off by much as 0.000375 volts.

INTERFACING DIGITAL CONTROLLER WITH ANALOG WORLD   115

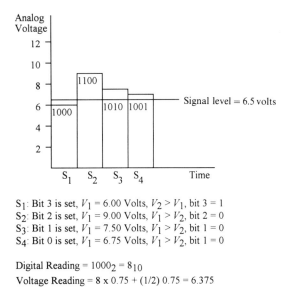

S_1: Bit 3 is set, $V_1 = 6.00$ Volts, $V_2 > V_1$, bit 3 = 1
S_2: Bit 2 is set, $V_1 = 9.00$ Volts, $V_1 > V_2$, bit 2 = 0
S_3: Bit 1 is set, $V_1 = 7.50$ Volts, $V_1 > V_2$, bit 1 = 0
S_4: Bit 0 is set, $V_1 = 6.75$ Volts, $V_1 > V_2$, bit 1 = 0

Digital Reading = $1000_2 = 8_{10}$
Voltage Reading = $8 \times 0.75 + (1/2)\, 0.75 = 6.375$

Fig. 3.18 Successive approximation.

Accuracy and linearity errors are in addition to resolution inaccuracies. All these factors should be considered when choosing an appropriate converter based on the application and its desired resolution and accuracy.

The engineer or scientist taking a measurement is ultimately interested in how closely the recorded data point represents the physical measurement taken from the process, e.g. temperature or position. Hence, there are other sources of inaccuracy that must be considered in the overall measurement system, as illustrated in Fig. 3.19a. The transducer supplies the output voltage proportional to the physical measurement. Manufacturers of transducers provide information on the linearity and accuracy of the output signal in relation to the physical measurement. Here we see an output signal in the range of 0–100 millivolts.

In order to use the full scale of the A/D converter this signal must be amplified, a form of signal conditioning. Here an amplification of 100 will match the full scale of the transducer to that of the A/D converter. Since the amplifier is constructed with imperfect components, it too will have built-in inaccuracies. Finally, the amplified signal is fed to the A/D converter, the inaccuracies of which we have already spoken about.

Figure 3.19b shows all of the scales we have just discussed in relation to the measurement problem. It is well to remember that the ultimate interest is in recording a measured value for the physical attribute being measured, e.g. temperature. Figure 3.19b is an idealization of the instrumentation between the physical process and the digital computer. In general, the accuracy of a measurement should consider the sum of all the sources of inaccuracy in the

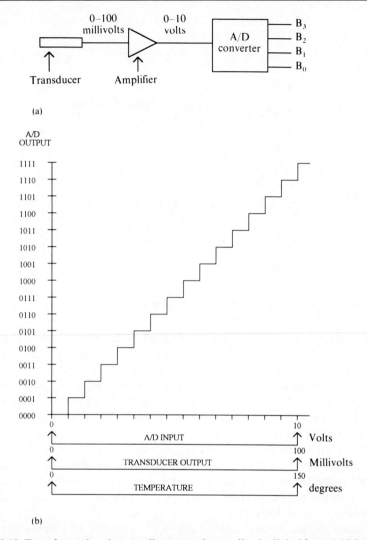

Fig. 3.19 Transformations in recording an analog reading in digital form. (a) Measurement system; (b) transformations among scales.

components of the measurement system, starting at the transducer/process interface and ending when a digital count is recorded in the computer.

3.5 TRANSDUCERS

Having examined the basic elements of a sensing system, we now focus our attention on the variety of transducers that are available for converting physical phenomena into electronic signals. A useful way to classify transducers is on the basis of the physical property the device is intended to measure. The important properties discussed in this section are:

- position
- velocity
- force or pressure
- temperature

3.5.1 Position transducers

Position transducers are widely used in servomotors, linear position tables, and other applications where precise positioning is important. In this section we will discuss four analog position transducers (potentiometers, linear variable differential transformers, floats and resolvers) and two digital position transducers (the optical encoder and ultrasonic range sensor).

(a) Potentiometers

The basic operation of a potentiometer was described in section 3.3.2 for a linear potentiometer. This device can be used, for example, in measuring linear travel of a sliding table. The characteristic function of a linear potentiometer is:

$$V_o = KXV_{in} \qquad (3.7)$$

where K is a constant and V_o, X and V_{in} are defined in Fig. 3.12.

Potentiometers are also made with rotary brushes, as shown in Fig. 3.20. This construction is useful for mounting on a rotary device, such as the rotor of a DC motor. In this case, angular position is related to output voltage by the equation:

$$V_o = K\theta V_{in}, \qquad (3.8)$$

where θ is the angular position of the wiper.

Potentiometers are quite inexpensive, very rugged, and easy to use. However, they are not as accurate as some other position transducers.

(b) Linear variable differential transformers

The linear variable differential transformer (LVDT) is a high-resolution contact transducer. As Fig. 3.21 illustrates, it is constructed with three coils, one primary and two secondary. A magnetic core sits within the coils.

If an alternating current is imposed on the primary coil, a voltage will be induced across the secondary coil. The magnitude of that voltage is a linear function of the position of the magentic core. Deviations from the null position

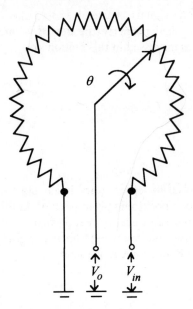

Fig. 3.20 Rotary potentiometer.

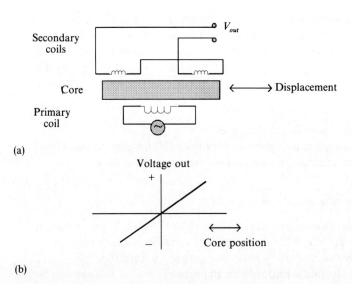

Fig. 3.21 Principle of a linear variable differential transformer. (a) Operation of a LVDT; (b) core displacement and voltage out.

of the core are translated into voltage readings by the equation:

$$\Delta V_o = K\Delta X, \qquad (3.9)$$

where

ΔV_o = change in output voltage
K = a proportionality constant
ΔX = change in position

LVDTs come in varying sizes. A typical limit of travel is in the range of 3 inches. The resolution of an LVDT is excellent, easily able to measure displacement below 0.001 inches. Since this is an analog device, the limits of resolution are usually governed by the resolution of the A/D converter.

LVDTs use alternating current. Therefore, a requirement exists to transform the output voltage to DC before it is applied to the A/D converter. Manufacturers of LVDTs service this requirement with instrumentation packages that provide the required DC output voltage.

It is readily apparent that the LVDT has an advantage over the potentiometer as a position measurement device. Since its core does not touch the coil, there is no mechanical wear that would result in deterioration of performance over time. On the other hand, it is a more expensive transducer, justifiable primarily where very high and repeatable accuracy is required.

(c) Resolvers

The resolver, shown in Fig. 3.22, is a rotary transformer. The primary winding is on the rotor and the secondary windings are on the stator. The secondary windings are set 90 degrees apart. An alternating current is imposed on the primary winding. As the shaft is rotated by the device where position is being monitored, the voltages on the secondary windings will vary as the sine and cosine of the angle of the rotor.

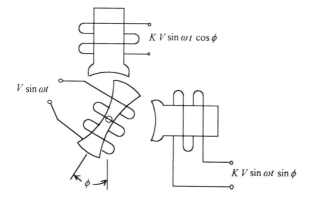

Fig. 3.22 Principle of a resolver.

The two output voltages can be converted to a binary count using a resolver-to-digital converter (RDC). This device combines the circuitry necessary to interpolate the output voltages into angular positions and the A/D circuitry required to digitize the result.

The resolver is a non-contact transducer. Unlike the rotary potentiometer, it will not lose accuracy due to wear. The high precision of the resolver is usually limited by the A/D converter, which converts the analog voltage into a digital count. Typical A/D converters incorporated in RDCs have 12- to 16-bit resolution.

(d) Optical encoders

An optical encoder is a digital position measuring device. It is available in both linear and rotary construction. Figure 3.23 shows the principle of an optical encoder in rotary form. A slotted disc is rotated in the path of a photo emitter/detector pair, in principle similar to that discussed in section 3.2.3. The emitter is typically a light-emitting diode and the detector is a photo-sensitive transistor. When light is incident on the base of the transistor, current flows from collector to emitter. As the disc rotates, the light will be alternatively blocked and allowed to pass. In the blocked state, the transistor will stop conducting. The voltage output from the detector circuit is a saw tooth. This is fed into a Schmitt trigger, which is a digital device that converts the saw tooth pattern into a square wave. The square wave, with amplitude of five volts, is digital data in a form that is readable by a digital controller. It is typical for the detector circuit to output a high signal when light is blocked and a low signal when a slot is encountered.

There are basically two implementations of the above principle in an optical encoder. The two types of implementations are the absolute encoder and the incremental encoder.

In an absolute encoder the disk is cut so that the code appearing along any radial line represents the absolute value of the position. Figure 3.24 shows examples of absolute encoder disks. White sections represent slots where light can pass from emitter to detector. Four emitter/detector pairs are lined up radially so that all four positions can be sensed simultaneously as the disk turns. Thus, if all four detectors are conducting at the same time, the disk is in

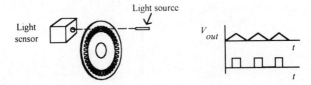

Fig. 3.23 Principle of an optical encoder.

TRANSDUCERS

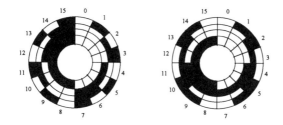

POSITION (decimal value)	BINARY	GREY
0	0000	0000
1	0001	0001
2	0010	0011
3	0011	0010
4	0100	0110
5	0101	0111
6	0110	0101
7	0111	0100
8	1000	1100
9	1001	1101
10	1010	1111
11	1011	1110
12	1100	1010
13	1101	1011
14	1110	1001
15	1111	1000

Fig. 3.24 Four-bit binary and grey code.

position 0. There are two kinds of codes used for an absolute encoder: binary code and grey code.

Figure 3.24 shows 4-bit binary code and 4-bit grey code juxtaposed. The binary code implementation uses the binary number system. A difficulty with using binary is that, as the encoder turns, there will be transitions in which two bits will change simultaneously. When reading multiple input signals at high speeds there is a risk of missing the transition on a bit. Grey code tries to eliminate this condition by using a pattern that allows only one bit to change at a time.

A high-resolution absolute encoder may have 10 to 15 concentric rings of binary code. This yields a relatively high angular resolution of 0.35 degrees to 0.01 degrees. High-resolution absolute encoders are relatively expensive. Incremental encoders are simpler and less expensive than absolute encoders, but they can only measure position relatively.

The construction of a typical incremental encoder is shown in Fig. 3.25. There are three photosensors used. Two photosensors are offset at the slots to indicate both the number of transitions and the direction. The third photosensor is a reference slot that indicates when one complete revolution has occurred.

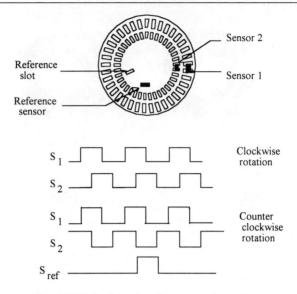

Fig. 3.25 Principle of an incremental encoder.

Figure 3.25 shows the data transmission for clockwise and counterclockwise rotation. If the disc is moving in a clockwise direction, the sensor $S1$ will always be activated first. If the disc is moving counterclockwise, the sensor $S2$ is always activated first. If the disc is rotated beyond 360 degrees, the reference slot will be encountered.

The incremental encoder is a relative position device. It keeps track of the shaft position relative to the reference point. It is subject to inaccuracy because it is possible that a transition will be missed in the count, leading to an incorrect intermediate position. The only absolute position is the reference point.

(e) Float transducers

Float transducers are the simplest method of measuring continuous position (height or level) of a liquid in a tank. Floats are widely used in process industries in which batches of liquids are feeding production processes. There are different implementations of a float; one simple device is shown in Fig. 3.26. Here a float is attached to a rod that moves the wiper of a rotary potentiometer. From equation 3.8, the angle of displacement is:

$$\theta = \frac{V_o}{KV_{in}}$$

The level, L, is related to θ as follows:

$$L = H - d\sin\theta. \qquad (3.10)$$

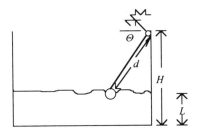

Fig. 3.26 Principle of a float transducer.

(f) Ultrasonic range sensors

For continuous measurement of a level in a tank, floats are being supplanted by ultrasonic range sensors, an example of which is shown in Fig. 3.27. Ultrasonic sensors use pulses of sound to measure distance. A transmitter sends out a pulse that is reflected against the fluid of which the level is being measured. When the transmitter sends out the pulse, it simultaneously initiates a timer circuit that counts clock cycles. A receiver, housed with the transmitter,

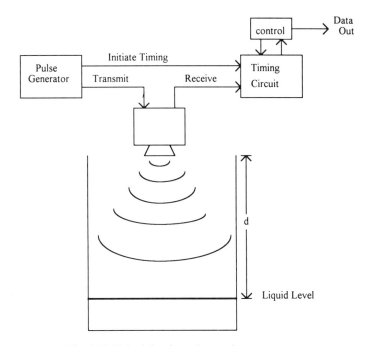

Fig. 3.27 Principle of an ultrasonic range sensor.

receives the reflection of the sound. The received signal terminates the timer and initiates a computation of distance. A microprocessor computes distance based on the speed of sound through the medium, typically air. The microprocessor may take several samples and compute an average to obtain a more accurate measurement.

The reflected signal will travel $2d$ during the period that the timer is on, Δt. If v is the velocity of sound in the medium, the distance between transducer and liquid level is

$$d = \frac{1}{2}\frac{v}{\Delta t}$$

3.5.2 Velocity and flow rate transducers

Velocity transducers are used for speed control. We shall describe the primary digital (optical encoder) and analog (DC tachometer) velocity transducers. When measuring the velocity of a fluid in a pipe, which is a parameter of interest in the process industries, the term **flow rate** is used. This section also describes some typical flow rate transducers.

(a) Optical encoders

In the previous section we described the optical encoder as a position transducer. Since velocity is the positional change with respect to time, any positional transducer can be used to measure velocity.

If θ_1 and θ_2 are two sequential angular positions of the encoder given in radians, then

$$\omega_i = \frac{\theta_2 - \theta_2}{\Delta t}, \tag{3.11}$$

where:

$\omega_i =$ the instantaneous angular velocity in radians/sec
$\Delta t =$ the increment of time between sequential position changes.

(b) Tachometers

The basic analog velocity measurement device is the tachometer, or generator. This device can be based on the operating principle of an AC generator or a DC generator. The principle of a DC machine will be covered in detail in the next chapter. As shown in Fig. 3.28, a DC tachometer consists of an armature (rotating conductor) mounted on the shaft of a device whose angular velocity is to be measured. The stator, or stationary component of the tachometer, is a permanent magnet. As the rotating conductor passes through the magnetic field, a current is induced in the conductor, resulting in a measurable voltage at

TRANSDUCERS

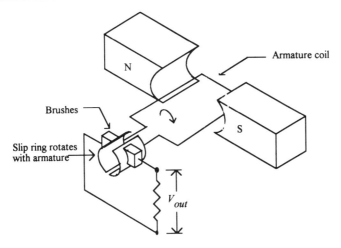

Fig. 3.28 Principle of the DC tachometer or DC generator.

V_{out}. In an ideal tachometer the relationship between speed and velocity is linear. Simply put,

$$\frac{V_{out}}{\omega_{in}} = K_\tau \qquad (3.12)$$

where K_τ is the tachometer constant. Since the tachometer is an analog device, A/D conversion is required.

(c) Flow rate transducers

The velocity, or flow rate, of a fluid in a pipe is a very important measure in the process industries. There are many different transducer technologies developed for this purpose; here we look at two common implementations.

Figure 3.29 shows a method based on the measurement of differential pressures. The pipe is purposely restricted at a point along the path of liquid flow. Based on the conservation of flow, the flow rate of fluid past any point in the pipe is equal throughout the pipe. In particular, the equation of continuity of fluid flow requires that

$$Q = A_1 v_1 = A_2 v_2, \qquad (3.13)$$

where:

Q = flow rate in in³/s, a constant throughout the pipe
A_1 and A_2 = cross-sectional area in in²
v_1 and v_2 = velocity of fluid flow in in/sec.

Since the cross-sectional area is smaller at the constriction, the velocity of fluid

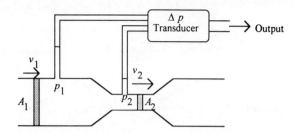

Fig. 3.29 Principle of a differential pressure flow transducer.

will be greater. This difference in fluid flow rate will result in a differential pressure between p_1 and p_2. The relationship between pressure and velocity is determined by Bernoulli's equation:

$$p_1 + (1/2)\rho v_1^2 = p_2 + (1/2)\rho v_2^2, \tag{3.14}$$

where:

p_1 and p_2 = pressure in lbs/in^2

ρ is the viscosity of the fluid.

Since the velocity is greater at the restriction, the pressure is smaller. The rate of fluid flow is measured by the pressure differential between the pipe and the restriction. Rewriting equation 3.14,

$$v_2^2 - v_1^2 = (2/\rho)(p_1 - p_2).$$

Substituting from equation 3.13,

$$Q^2 = A_2^2 v_2^2 = \frac{2\Delta p}{\rho} \left(\frac{A_1^2 A_2^2}{A_2^2 - A_1^2} \right)$$

$$Q = K\sqrt{\Delta p} \tag{3.15}$$

where:

K = a constant

Δp = the pressure differential.

Equation 3.15 shows that Q is proportional to the square root of Δp. Therefore, implementation of this principle requires a transducer that will output a linear signal from the square root of differential pressure.

Another common method for measuring flow is the turbine flowmeter, illustrated in Fig. 3.30. The turbine is mounted on a rotating shaft and the flow rate of the fluid causes rotation of the turbine in proportion to the rate of fluid flow. One method of measuring velocity of the turbine rotation is to use a magnetic pick-up coil mounted external to the pipe. As the turbine rotates, an AC voltage is induced in the pick-up coil. Each time a blade passes the coil,

TRANSDUCERS

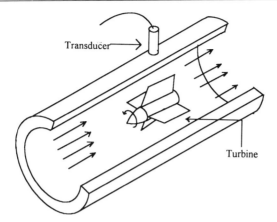

Fig. 3.30 Principle of a turbine flowmeter.

a voltage pulse is induced. By measuring the number of pulses per unit time, the angular velocity of the turbine and, hence, the rate of flow of the fluid can be measured. An alternative configuration is to mount a sealed tachometer on the turbine shaft and measure flow rate as a function of the voltage output of the tachometer.

3.5.3 Force or pressure transducers

Force sensors are used extensively in automatic weighing operation in the process industries and in robotic applications when it is necessary to control gripping pressure. In this section we shall examine two analog transducers: the load cell and the strain gage.

(a) Load cells

A load cell is used in processes where precise weighing is required. It can be implemented using a strain gage or an LVDT. Figure 3.31 illustrates a load cell implementated by using a LVDT and a spring with linear force displacement. The appropriate transfer function is

$$F = K\Delta d \qquad (3.16)$$

where:

$F = $ force
$K = $ the spring constant
$\Delta d = $ displacement from the unloaded rest position.

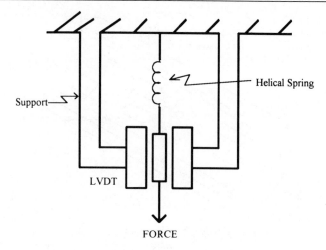

Fig. 3.31 LVDT used in weight measurement.

(b) Strain gages

The most widely used pressure and force sensitive transducer is the strain gage. The principle of the strain gage is based on the resistance properties of electrical conductors. Electrical conductors possess resistance based on the relationship:

$$R = \rho(L/A) \qquad (3.17)$$

where:

R = resistance in ohms
ρ = the resistivity constant, which is a property of the specific conductor material being used in the wire. It is measured in units of ohm-cm
L = the length of the wire in cm
A = the cross-sectional area in cm^2

Since the resistivity is a constant, a change in the length and/or area of the wire will cause a change in the resistance. This phenomenon is called 'piezoresistivity'.

Figure 3.32 shows the influence of forces in tension on a conductor. At the top of the figure, the conductor is at rest. At the bottom of the figure, the conductor is in tension, increasing its length and reducing its area. The resistance of the strain gage changes in proportion to its changing dimensions.

$$R_0 + \Delta R = \rho \frac{L_0 + \Delta L}{A_0 - \Delta A}$$

where

$\Delta R, \Delta L,$ and ΔA = the change in resistance, length and area, respectively
$R_0, L_0,$ and A_0 = the resistance, length, and cross-sectional area at rest.

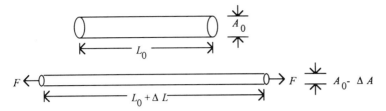

Fig. 3.32 Principle of a strain gage.

The gage factor, G, of a strain gage is the ratio of the relative change in resistance to the relative change in length.

$$G = \frac{\Delta R/R_0}{\Delta L/L_0}$$

These are two primary constructions used in making strain gages: bonded and unbonded. These are shown in Fig. 3.33. In the unbonded strain gage the wire resistance element is stretched taut between two flexible supports. The wire stretches in accordance with the forces applied to the diaphragm. The resistance of the wire changes in relation to these forces.

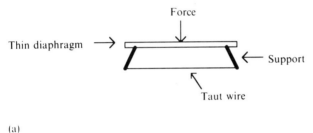

(a)

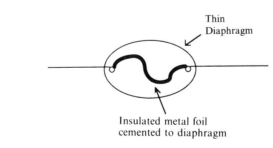

(b)

Fig. 3.33 Strain gage construction. (a) Unbonded; (b) bonded.

In a bonded strain gage, a wire or a metal foil is placed in a thin metal diaphragm. When the diaphragm is flexed, the element deforms and a change in resistance occurs. In general, bonded strain gage construction is more durable than unbonded.

3.5.4 Temperature tranducers

Temperature transducers are used extensively in process industries such as chemical, food and pharmaceuticals, where control of temperature during manufacture is important. Three commonly used temperature transducers are the thermocouple, the resistance–temperature detector (RTD) and the thermistor.

(a) Thermocouples

When a temperature differential exists across the length of a metal, a small voltage differential will exist due to the migration of electrons in the metal. By joining two dissimilar metal wires together at one end, a small current will be induced at the junction due to differences in the molecular structure of the metals. This is shown in Fig. 3.34. Point (a), the juncture, is exposed to a temperature to be measured. Point (b), the other ends of the metal wires, is held at a reference temperature.

For dissimilar metals at a given temperature, the density of free electrons are different. This results in an electron migration at junction (a), causing a small current to flow from one metal to the other. This small induced electric differential, with proper signal conditioning, is measured at point (c). The electric signal has the property of being linear with the temperature differential

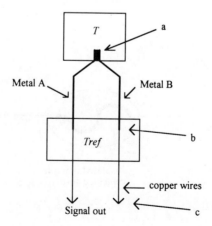

Fig. 3.34 Principle of a thermocouple.

between points (a) and (b). In particular, a simplified model is of the form:

$$V_{out} = \alpha(T - T_{Ref}) \quad (3.18)$$

where:

V_{out} = induced voltage
α = a constant in volts/degree K
T, T_{Ref} = the measured temperature and reference temperature, respectively.

(b) Resistance–temperature detectors (RTD)

The RTD temperature sensor is based on a particular property of metals wherein their electrical resistance changes with temperature. In particular, as temperature increases, so does electrical resistance. This is due to the fact that a higher temperature in a metal results in electron vibrations that impede the flow of free electrons in the metal.

When discussing strain gages in section 3.5.3, we introduced the relationship

$$R = \rho(L/A) \quad (3.19)$$

where:

R = resistance
ρ = resistivity
L = length of the wire
A = cross-sectional data.

In fact, ρ is a temperature sensitive parameter. With knowledge of the relationship between ρ and T for a particular metal, it is possible to construct an RTD with a specific relationship between R and T.

(c) Thermistors

A thermistor is made of a semiconductor material that exhibits a predictable and repeatable change in resistance as temperature is changed. Unlike a metal, the molecular structure of a semiconductor is such that increasing its temperature reduces its resistance. As the temperature of the material increases, electrons break free of their covalent bonds and conductivity is improved. The response function is as follows:

$$R_t = R_0 e^{\beta[(1/T) - (1/T_0)]} \quad (3.20)$$

where:

R_t and R_0 = the resistance of the thermistor at temperature t and at a reference temperature, respectively
T and T_0 = the thermistor temperature and the reference temperature, respectively, in degrees Kelvin

β = a property of the material used to make the thermistor, in units of degrees Kelvin
e = the base of the natural log

The non-linearity of the thermistor response function makes its use limited only over the most linear range of the device. Manufacturers of thermistors specify the useful range and the percent error over that range.

3.6 SUMMARY

Sensors can be broadly divided into two types: discrete and continuous. Discrete sensors are used in applications for which it is only necessary to know the state of the physical process being sensed. Continuous sensors are used when it is necessary to measure the magnitude of some physical property of the process.

In this chapter we examined the components of a sensing system, which includes a sensor and measurement circuit, and may include some form of signal processing. Sensors, their measurement circuit and, sometimes, a signal conditioner are designed and sold by vendors as a complete package. The term 'transducer' is used to describe such a sensing system. When the electrical output of a transducer is to be interfaced to a computer, an analog to digital converter is required. The functioning of A/D and D/A converters were discussed in some detail and a number of typical transducers were described for measuring position, velocity, force and temperature.

EXERCISES

1. A manufacturer produces two types of parts, one made of metal and one made of plastic. The machines that produce the parts dump them on a main conveyor. The company wishes to develop a sorting system that will sense which part is coming down the main conveyor and divert the part down the appropriate lane for packaging. It has been suggested that non-contact sensors be used to sense the part before it reaches the diverter lane. After a time delay, the diverter could be energized to push the part down the diverter lane. The diverter rods have limit switches that are enabled when they reach the end of their stroke. The situation is shown in Fig. 3.35. You have been asked to participate in the design of the system.

    (a) Which lane would you use to divert the plastic part and which sensors would you use? The metal part?
    (b) Assuming you are using a controller with parallel ports B and C as defined in Fig. 2.37, define all the inputs and outputs you would need to control the system.

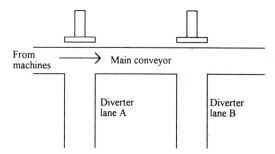

Fig. 3.35 System for Exercise 1.

(c) Write a control program to operate the system. Assume a time delay of 0.5 seconds between sensing the part and starting the extension of the diverter. The controller clock speed is 2 MHz.

2. A rotary potentiometer is used as a position sensor. The maximum angular displacement to be measured is 180 degrees, and the potentiometer is rated for 10 volts over a range of 270 degrees of rotation.

   (a) What voltage increment must be resolved by an A/D converter in order to measure angular displacement of 0.75 degrees?
   (b) How many bits are required by the A/D converter for detecting incremental displacement of 0.75 degrees?

3. For the situation in Exercise 2, assume that the potentiometer is replaced by an encoder with 360 degree range.

   (a) How many slots are required on an incremental encoder in order to obtain the resolution of 0.75 degrees?
   (b) How many circular tracks are required on an absolute encoder in order to obtain the same resolution?

4. A temperature transducer has been instrumented to give an output of 0–10 millivolts over a temperature range of 50 to 200 degrees. The reading is being made by a controller with an 8-bit A/D converter and a 10-millivolt reference.

   (a) If the probe is emitting 5.35 millivolts, what digital reading is being logged by the controller?
   (b) What voltage is this interpreted as?
   (c) What temperature is this interpreted as?

5. An 8-bit D/A converter with a 10-volt reference has an input of 10010001.

   (a) What is the output voltage?
   (b) What is the step size of the D/A converter?
   (c) What is the percent resolution?

6. The D/A converter of Exercise 5 has $R_f = R$. What happens to the percent resolution and step size if R_f is changed to $2R$?
7. A transducer is being calibrated to measure distance over the range 0–4 mm. The results of the calibration are shown in the table below.

Calibration data	
(mm)	(millivolts)
0.0	0.0000
0.5	1.1204
1.0	2.2343
1.5	3.3418
2.0	4.4429
2.5	5.5375
3.0	6.6257
3.5	7.7075
4.0	9.0000

(a) What should the data point in millivolts be for an 'ideal' calibration?
(b) I am going to amplify the signal to a range of 0–9 volts and capture the data using an A/D converter calibrated over the range 0–10 volts. What digital value will I record in the computer at 3.5 mm? How does this compare with the ideal digital count?

8. A tachometer is required to measure the velocity of a rotary actuator over the range 0–3000 rpm. There are two tachometers available with constants of 0.025 V/rad/s and 0.021 V/rad/s. Data will be collected using an 8-bit A/D converter with a reference voltage of 10 volts. Assuming that both tachometers have the same linearity, decide which tachometer to use.

9. Two parameters given by manufacturers of tachometers are the tach constant and the percent linearity over the full range of the tachometer. The tach constant is the average of the sums of output voltage in volts to speed, either in rpm or rad/s, over selected test speeds within the full range of operation. The percent linearity is found by calculating a theoretical output voltage for these test speeds using the computed tach constant. The percent linearity is the worst case deviation from the theoretical line over the entire range of operation and can be computed from the formula:

$$\% \text{ linearity} = \frac{V_{actual} - V_{theoretical}}{V_{theoretical}} \times 100$$

A test fixture was used to collect data on a particular model of tachometer, which is to be used to collect data over the range 1000–4000 rpm. What is the tachometer constant and the percent linearity?

rpm	Test data actual output voltage
1000	3.00
2000	6.10
3000	9.10
4000	12.08

10. A strain gage wire 0.02 mm in radius and 5 cm long has a resistance of 40 ohms when at rest and 45 ohms when stretched to 5.3 cm. It is to be used in tension over that range.

   (a) What is the resistivity of the material?
   (b) Assume I placed the strain gage in a half bridge circuit as shown in Fig. 3.36. What will be the range of V_{out} over the full range of the strain gage?
   (c) A commonly used circuit for strain gage applications is the full bridge, of which the Wheatstone bridge (Fig. 3.37) is an example. This bridge allows the user to null the output voltage to zero when the gage is at rest by choosing appropriate values of R_1, R_2, and R_3. Given the values of 40 ohms for each resistor, compute the full range of V_{out} over the full range of the strain gage?

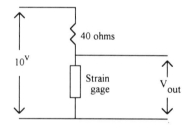

Fig. 3.36 Strain gage in a half-bridge.

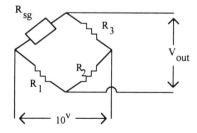

Fig. 3.37 Strain gage in a Wheatstone bridge.

FURTHER READING

Banner Engineering Corporation (1992) *Photoelectric Controls*, Banner Engineering Corporation, Minneapolis, Minnesota.

Considine, D. M. (ed.) (1974) *Process Instruments and Controls Handbook*, McGraw-Hill Book Company, New York, New York.

de Silva, C. W. (1989) *Control Sensors and Actuators*, Prentice-Hall, Inc, Englewood Cliffs, New Jersey.

Goldsbrough, P. F., Lund, T. and Rayner, J. P. (1983) *Analog Electronic for Microcomputer Systems*, The Blacksburg Group, Inc., Blacksburg, Virginia.

Johnson, C. D. (1988) *Process Control Instrumentation Technology*, 3rd edn, John Wiley & Sons, Inc., New York, New York.

Nechtigal, C. L. (1990) *Instrumentation and Control: Fundamentals and Applications*, John Wiley & Sons, Inc., New York, New York.

Norton, H. N. (1982) *Sensor and Analyzer Handbook*, Prentice Hall, Inc., Englewood Cliffs, New Jersey.

Parr, E. A. (1987) *Industrial Control Handbook Volume 1: Transducers*, Industrial Press, Inc., New York, New York.

Tocci, R. J. (1980) *Digital Systems: Principles and Applications*, Prentice-Hall, Inc., Englewood Cliffs, New Jersey.

Actuators and the performance of work 4

4.1 INTRODUCTION

In Chapter 1 we described the general concepts of open and closed loop electromechanical systems for performing work. The muscle for performing the work is provided by an actuator, which is typically a motor or other device that converts electrical input into mechanical action. This chapter will describe the operation and use of typical actuators, which include DC motors and servomotors, stepping motors, relays and solenoids, and pneumatic power systems. Specifying an actuator for a specific operation requires measuring the work to be performed in order to specify the power and torque/speed requirements of the actuator that will perform the work. Thus, the description of actuators will be preceded by a review of some basic concepts of physics as they apply to electromechanical systems.

4.2 CONCEPTS OF WORK, FORCE, TORQUE AND POWER

Systems engineered to perform work typically combine a prime mover, such as an electric motor, with a drive system, such as a combination of mechanical levers, gears and pulleys. Figures 4.1a and 4.1b illustrate two simple systems: a motor driving a rack through a gear and a motor driving a pulley. Useful work is performed at the end of the rack and pulley by moving a load through a distance.

It is conceptually useful to understand that the description of work performed in electromechanical systems can be broken down and measured in terms of linear action and rotary action. In specifying an actuator for a particular application, the problem is to analyze the requirements, such as speed and force at the work/machine interface, and to specify the actuator capable of delivering those requirements.

The definition of mechanical **work** is the application of a force over a distance. For linear movement over a distance S:

$$W = F \cdot S \tag{4.1}$$

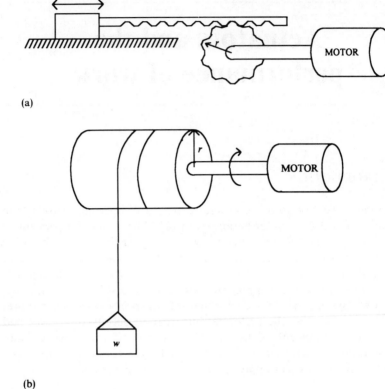

(a)

(b)

Fig. 4.1 Examples of prime mover and drive system. (a) Rack and pinion moving a load; (b) pulley lifting a weight.

where:

W = work, e.g. in ft lbs
F = force, e.g. in lb
S = distance, for example, in ft

The **power** being expended by providing a constant force over a linear distance during an interval of time is defined as:

$$P = \frac{dW}{dt} = \frac{d(F \times S)}{dt} = F\frac{dS}{dt} = F \times v \qquad (4.2)$$

where:

P = power, e.g. in lb ft/s
v = velocity, e.g. in ft/s

The dual of a linear system is rotary, which is the motor and gear of Fig. 4.1a and the motor and drum of Fig. 4.1b. Applying equation 4.1, the work

CONCEPTS OF WORK, FORCE, TORQUE AND POWER

performed in one complete revolution of the rotary member is:

$$W = F \cdot S = F(2\pi r), \tag{4.3}$$

where:

r = radius, or distance of force from center of rotation
2π = the number of radians in one complete revolution

In general,

$$W = F\theta r, \tag{4.4}$$

where:

θ = the distance rotated in radians.

EXAMPLE 4.1

How much work is performed in moving a 10 lb weight vertically upward 2 ft as shown in Fig. 4.1b? The radius of the drum is 6 in. Assume a frictionless system with no inertia in the motor or drum.

Answer

The force exerted by a body in equilibrium is its weight. From the point of view of the linear displacement of the weight,

$$W = F \cdot S = (10\,\text{lb})(2\,\text{ft}) = 20\,\text{ft lb}.$$

From the point of view of the rotary system, the angular movement required is:

$$\theta = \frac{\text{Linear distance}}{\text{Circumference}} \times 2\pi$$

$$\theta = \frac{2\,\text{ft}}{2\pi r} \times 2\pi = 4\,\text{radians}$$

$$W = F\theta r = (10)(4)(0.5) = 20\,\text{ft lb}.$$

Note that the work requirements can be computed from either the equations of the linear model or the equations of its dual rotary model.

The work required by the mechanical system must be provided by the prime mover, usually a motor. An important measure of the performance capability of a rotating machine is **torque**, the twisting movement producing rotation.

$$T = F \cdot r \tag{4.5}$$

where:

T = torque, e.g. in lb ft.

Therefore,
$$W = T \cdot \theta \tag{4.6}$$

For a rotating machine, power can now be defined as:

$$P = \frac{dW}{dt} = \frac{d(T \times \theta)}{dt} = T\frac{d\theta}{dt} = T\omega \tag{4.7}$$

where:

ω = the angular velocity in rad/s.

EXAMPLE 4.2

How much torque can be delivered by a 3 HP (horsepower) motor running at 100 rpm? At 200 rpm?

Answer

$$1\,\text{HP} = 550\,\text{ft lb/s}.$$

At 100 rpm:

$$T = \frac{P}{\omega} = \frac{(3)(550)\,\text{ft lb/s}}{(100 \times 2\pi/60)\,\text{rad/s}} = 158\,\text{lb ft}$$

At 200 rpm:

$$T = \frac{(3)(550)(60)}{200 \times 2\pi} = 79\,\text{lb ft}$$

Torque is an important design parameter of rotary actuators, since they must be specified so that they can exert specific forces at a distance from their shafts in order to perform the required work. Moreover, the rotary actuator must be chosen for the worst case (maximum) torque that will be required. In addition, as illustrated in Example 4.2, there is an inverse relationship between torque and angular velocity that must be considered. In general, a DC motor can exert higher torques at lower velocity.

EXAMPLE 4.3

A motor is required to lift a weight at a constant velocity as shown in Fig. 4.1b. The weight is 10 lb and the desired constant velocity is 5 ft/s. The drum has a radius of 6 in. Assuming a frictionless system and no inertia, what horsepower motor is required?

CONCEPTS OF WORK, FORCE, TORQUE AND POWER

Answer

The torque is a force exerted at a distance, in this case 10 lbs at a distance of 6 in.

$$T = F \cdot r = (10)(1/2) = 5 \text{ lb ft.}$$

Since this torque is to be delivered at a constant velocity of 5 ft/s,

$$P = T\omega = (5 \text{ lb ft}) \frac{5 \text{ ft/s}}{(2\pi)(0.5) \text{ ft}} (2\pi) = 50 \text{ lb ft/s}$$

$$P = \frac{50}{550} = 0.09 \text{ HP}$$

Example 4.3 is very simplified. In reality, torque is required to overcome two loads on the rotary machine. The first arises from the work that must be done on the process. This work may result from lifting a load, as in Example 4.3, or in overcoming friction, as in the case of the cutting tool of a milling machine, or in the gears of the drive system. This is usually called the run torque and it exists whenever work is being done. The second load is due to inertia, which arises during acceleration of a body. **Inertia**, which is the tendency of a body to remain at rest or in a constant velocity and direction, can arise from the rotating parts of the motor itself, the workload, or the drive system between the motor and the workload. This is often called start-up torque, since it occurs at startup during acceleration to working velocity.

The relationship between torque and inertia for an accelerating body can be developed from the basic laws of motion. Newton's equation is:

$$F = \frac{w}{g} \frac{d^2 s}{dt^2} = ma \tag{4.8}$$

where:

- w = the weight of the body (lb)
- g = gravitational acceleration (32.2 ft/s^2)
- $m = w/g$, the mass of the body (a constant in any gravitational field)
- a = acceleration (ft/s^2)

Since $T = Fr$,

$$T = mra.$$

The instantaneous acceleration in a linear system is

$$a = \frac{v_2 - v_1}{\Delta t}.$$

Therefore,

$$T = mr\left(\frac{v_2 - v_1}{\Delta t}\right)$$

where $v_2 - v_1$ is the change in velocity over the interval Δt. In a rotary system, $v = r\omega$ and

$$T = mr^2\left(\frac{\omega_2 - \omega_1}{\Delta t}\right) = mr^2\alpha \qquad (4.9)$$

where α is the angular acceleration.

For a rotating particle, the moment of Inertia, I, is defined as $I = mr^2$. Substituting into equation (4.9), we obtain the relationship between torque and inertia:

$$T = I\alpha \qquad (4.10)$$

The moment of inertia of a rotating body is the sum of the inertia of its particles, or

$$I = \sum_p m_p r_p^2$$

where P denotes particle. Some useful approximations for common geometries are given in Fig. 4.2. In the case of the thin walled cylinder, where the entire mass is rotating at a distance r from the center of rotation,

$$I = \sum_p m_p r_p^2 = mr^2 = \frac{wr^2}{(32.2\,\text{ft/s}^2)}$$

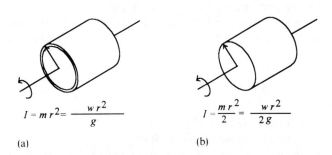

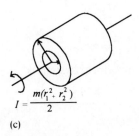

Fig. 4.2 Moment of inertia for some common bodies. (a) Thin-walled cylinder; (b) solid cylinder; (c) thick-walled cylinder.

CONCEPTS OF WORK, FORCE, TORQUE AND POWER

In the case of the solid cylinder, the average particle is a distance $r/2$ from the center of rotation, and

$$I = \frac{mr^2}{2} = \frac{wr^2}{2(32.2 \text{ ft/s}^2)}$$

EXAMPLE 4.4

What is the torque required by the motor in Fig. 4.1b to lift a weight of 10 lb? Assume the drum is a thin walled cylinder having a radius of 6 in and weights 4 lb. The desired velocity during lift is 15 ft/s and the time to reach velocity is 0.5 s. Manufacturers specifications show that the inertia of the motor rotor is 0.2 lb-ft-s².

Answer

Startup torque is required to (1) accelerate the weight, (2) accelerate the drum, and (3) accelerate the motor rotor. Run torque is required to lift the weight at terminal velocity.

1. Inertia of the weight:

$$m_w = \frac{w}{g} = \frac{10 \text{ lb}}{32.2 \text{ ft/s}^2} = 0.31 \text{ lb/ft/s}^2$$

$$I_w = m_w r^2 = (0.31)(0.5)^2 = 0.0775 \text{ lb-ft-s}^2$$

2. Inertia of the drum:

$$m_d = \frac{w}{g} = \frac{4 \text{ lb}}{32.2 \text{ ft/s}^2} = 0.123 \text{ lb/ft/s}^2$$

$$I_d = m_d r^2 = (0.123)(0.5)^2 = 0.031 \text{ lb-ft-s}^2$$

3. Inertia of the motor rotor:

$$I_r = 0.02 \text{ lb-ft-s}^2$$

4. Total inertia:

$$I = 0.0775 + 0.031 + 0.02 = 0.1285 \text{ lb-ft-s}^2$$

5. Angular acceleration:

$$\Delta\omega = \frac{\Delta v}{r} = \frac{15 \text{ ft/s}}{0.5 \text{ ft}} = 30 \text{ rad/s}$$

$$\alpha = \frac{\Delta\omega}{\Delta t} = \frac{30}{0.5} = 60 \text{ rad/s}^2$$

6. Torque required to overcome inertia:

$$T = I\alpha = (0.1285)(60) = 7.71 \, \text{lb ft}$$

7. Torque required to lift weight at terminal velocity:

$$T = Fr = (10 \, \text{lbs})(0.5 \, \text{ft}) = 5 \, \text{lb ft}$$

8. Maximum torque required:

$$T_{\text{total}} = 7.71 + 5.0 = 12.71 \, \text{lb ft}$$

Another common source of resistance to motion in a mechanical system is **viscous damping**. Viscous damping arises from housing a moving part in a bath of fluid or lubricant. This is often done as a way of reducing settling time when the moving part comes to rest. Viscous damping develops a torque that is proportional to the velocity of the moving part; i.e., $T = f\omega$, where f is the amount of viscous damping. If we combine torque from all sources,

$$\text{Total torque} = T_L + f\omega + I\alpha, \tag{4.11}$$

where T_L = the torque of the workload that is not proportional to velocity or acceleration.

In Example 4.4 it was specified that the load would be lifted at a certain velocity, but there was no specification of the total distance of the lift. In most situations, the total distance of the move and the time allowed to complete the move are important considerations when computing the torque requirements of a motor. For example, the requirement of Example 4.4 could have been to move the weight vertically a total of 60 feet in 8 seconds, with a maximum velocity of 15 ft/s. For this case, the problem requires the determination of a period of acceleration, followed by a period of constant velocity, followed by a period of deceleration.

Figure 4.3 shows a **move profile**, which is a graph commonly used to analyze the acceleration requirements of a move before computing the required torque of the motor. The move profile is trapezoidal. The horizontal axis represents

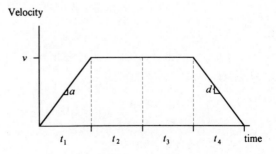

Fig. 4.3 Example of a move profile.

the total time of the move and the vertical axis represents velocity, where v is the maximum velocity. A good starting point is to divide the horizontal axis into four equal segments, where t_1 is the period of acceleration, t_4 is the period of deceleration, and $t_2 + t_3$ is the period of constant velocity. Therefore, the slope $a = v t_1^{-1}$ is the acceleration rate and $d = v t_4^{-1}$ is the deceleration rate. With Fig. 4.3 as a starting point, it is possible to make tradeoffs within the move profile while still achieving the goal of moving a specific distance in a specific time. For example, by shortening the period of acceleration and increasing the acceleration rate, it is possible to reduce the required maximum velocity. By increasing maximum velocity, it is possible to reduce the rate of acceleration.

4.3 POWER TRANSMISSION AND REFLECTED FORCES

The analysis of motor torque requirements arising from the mechanical system must take into consideration the effects of power transmission through drive systems such as gear trains, belts, and lead screws. Figure 4.4 illustrates the case for power transmission through gears. Here the position of a rack is being controlled by a motor directly driving a small gear which, in turn, drives a larger gear. In considering the motor requirements to move the load, the terms **reflected torque** and **reflected inertia** are used to indicate that the inertia and torque as seen by the motor is reflected through the intermediate gear.

With respect to the load only, F_1 is the force on gear 1 and the torque of the load about gear 1's center of rotation, $T_L = F_1 r_1$. Similarly, for gear 2, the torque about the motor shaft is $T_M = F_2 r_2$. By the laws of conservation of forces, $F_1 = F_2$ when friction in the gear train is ignored. Therefore, this leads to the conclusion:

$$\frac{T_M}{r_2} = \frac{T_L}{r_1}. \tag{4.12}$$

If we define $T_{\text{ref}} = T_M =$ the reflected torque of the load at the motor and $G = r_1/r_2$, then equation 4.12 can be rewritten:

$$T_{\text{ref}} = \frac{T_L}{G}, \tag{4.13}$$

where G is called the gear ratio. Note that the effect of the gear train is to reduce the torque of the load as seen by the motor.

It is clear from Fig. 4.4 that gear 1 will rotate at a slower angular velocity than gear 2. The relationship between motor speed and that of the drive gear is given by the gear ratio as follows:

$$\omega_1 = \frac{\omega_M}{G},$$

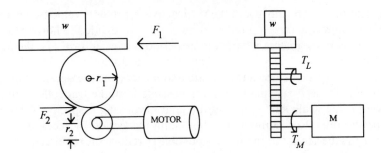

Fig. 4.4 Reflected torque and inertia in gear trains.

where ω_1 and ω_M are the angular velocities of gear 1 and motor shaft, respectively.

Since $T = I\alpha$ and using equation 4.12,

$$\frac{I_{ref}\alpha_M}{r_2} = \frac{I_L\alpha_1}{r_1},$$

$$I_{ref} = I_L\left(\frac{\alpha_1}{\alpha_M}\right)\left(\frac{r_2}{r_1}\right).$$

It can be shown that $\alpha_M/\alpha_1 = r_1/r_2 = G$. Therefore,

$$I_{ref} = \frac{I_L}{G^2}. \tag{4.14}$$

With appropriate gearing, the effects of inertia and torque of the load can be reduced, allowing a smaller motor to be used in an application. However, the gear train adds additional frictional forces and inertia from the gears themselves and this must be considered in the overall requirements of the system. This kind of analysis also applies to belt and chain driven systems. This analysis can also be extended to systems of *n*-coupled gears. This is left as an exercise for the reader at the end of the chapter.

Another important power transmission system in mechanical devices is the lead screw, which is shown in Fig. 4.5. The lead screw directly converts rotary motion into linear motion and is commonly found in positioning devices requiring precision. The screw is threaded and, in conventional lead screws, a nut rides on the thread and carries the load. Conventional lead screws are somewhat inefficient due to frictional forces arising from contact between the surfaces of screw and nut threads. This is considerably reduced in ball-bearing or ballnut lead screws, where ball bearings provide a rolling contact between the screw and nut.

The precision of a lead screw driven system is determined by the precision of the motor control and the lead screw pitch. The **lead screw pitch** can be defined

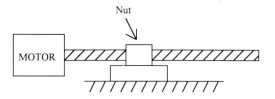

Fig. 4.5 Principle of a lead screw.

as the amount of linear travel of the nut for one complete revolution of the screw or, alternatively, the number of revolutions of the screw per length of linear travel. Using the latter concept,

$$P = \text{no. of threads per inch},$$

where P is the lead screw pitch in revolutions per inch. The nut moves the linear distance between 1 thread $(1/P)$ in one revolution.

The torque from a lead screw drive is greatest during the beginning of a move, when the inertial loads are being accelerated and coulomb friction is being overcome. The torque required to overcome axial forces due to friction can be computed as:

$$T_F = \frac{F}{2\pi P e}, \qquad (4.15)$$

where:

T_F = torque due to frictional forces
F = frictional forces, e.g. in ounces or lbs
e = lead screw efficiency in percent

Ball-bearing lead screws have efficiencies in the area of 90% while conventional lead screws can have efficiencies below 50%.

Calculating the inertial forces of the weight can be done using the relationship:

$$I = \frac{W}{g}\frac{1}{P^2}\left(\frac{1}{2\pi}\right)^2, \qquad (4.16)$$

where:

W = weight
g = acceleration due to gravity

In many cases it is not practical to compute the mechanical torque requirements of a machine to which a motor needs to be sized. For example, you may not have any data on an old conveyor from which the motor has been removed and you want to put it back into service. A practical way of obtaining an estimate of the power requirements is to use either a torque wrench or a scale. A torque wrench is used like an ordinary pipe wrench, but has a scale on it that provides a reading in ounce-inches. By attaching the torque wrench to the shaft that will be driven by the motor and using the wrench to turn the shaft, the torque required to start the machine rotating can be determined.

Another practical method is to use a weigh scale. Here you wrap a cord around the pulley or shaft that is to be driven. The other end of the cord is attached to a spring weigh scale. By pulling the spring scale until the shaft begins rotating and observing the maximum force registered on the scale, the formula $T = F \cdot r$, where r is the radius of the shaft, can be used to compute the required torque.

4.4 BASIC PRINCIPLES OF A DC MACHINE

In the previous section we evaluated the torque requirements to perform a specific function. The prime mover, or motor, must deliver that torque. In this section we examine the manner in which the torque is generated by the motor. A torque is generated by a DC motor due to the electromagnetic induction of force fields based on the effects of a magnetic field on a current-carrying conductor.

DC motors vary in the methods used to develop a magnetic field. In permanent magnet DC motors, a permanent magnet is used. Alternatively, an electromagnet can generate a magnetic field by wrapping a conductor around a medium that can be magnetized when current is passed through the conductor.

Figure 4.6 shows a current-carrying coil wrapped in air. The passage of current through the coil creates a field of magnetic flux, φ, about the coil.

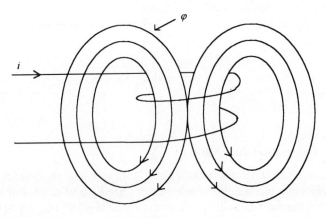

Fig. 4.6 Inducing magnetic flux with a current carrying coil.

BASIC PRINCIPLES OF A DC MACHINE

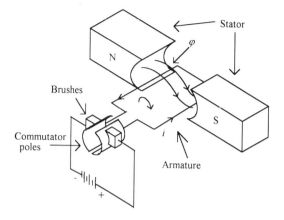

Fig. 4.7 Principle of a DC motor.

Magnetic flux is a measure of how much magnetism there is in the space about the coil. The direction of the flux can be determined by placing your right hand on the coil with your thumb pointing in the direction of current flow. Your fingers will then be pointing in the direction of the magnetic flux.

When an easily magnetized material, such as a permeable iron core, is placed within the coil wrap, the material will exhibit the properties of a magnet. The flux through the electromagnet will depend on the flux generated by the current-carrying conductor and the permeability of the medium (the core). Let

μ = the relative permeability of the medium
φ = magnetic flux in lines or webers

Then

$$\varphi_m = \mu\varphi,$$

where φ_m is the magnetic flux of the electromagnet. A typical ferromagnetic material used as the core of an electromagnet has a relative permeability of several hundred.

Whether a permanent magnet or electromagnet is used, it is the flux field of the magnet that is responsible for inducing motion in the machine. Figure 4.7 illustrates the basic arrangement of moving and stationary parts of a DC motor. The magnetic flux comes from the stationary part of the motor, called the stator. The rotating component of the motor, called the armature, is wrapped by coils of current-carrying wire.

If a current-carrying conductor is placed in a magnetic field, a force will act upon it. For the single conductor in Fig. 4.8, the force can be computed as

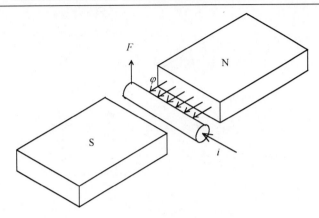

Fig. 4.8 Forces on a current-carrying conductor in a magnetic field.

follows:

$$F = BLi \qquad (4.17)$$

where:

B = magnetic flux density
L = length of the conductor in the magnetic field
i = magnitude of the current flowing in the conductor.

The **magnetic flux density** is a function of the magnetic flux and the area of the wire.

$$B = \frac{\varphi}{A} \qquad (4.18)$$

where:

A = the surface area of the conductor in the magnetic field.

The direction of force on the conductor is a function of the direction of the current and the magnetic field vectors. Assuming a conductor of unit length,

$$|F| = |i||B|\sin\theta$$

where θ is the angle between i and B. Since i and B form a plane, the direction of F will always be perpendicular to the plane, as shown in Fig. 4.8. When i and B are at 90 degrees, the force is maximum.

The dual of inducing a force vector with current in a conductor is to induce a voltage across a conductor by applying a force. This is the principle of a generator. If a conductor is moved through a magnetic field, a voltage appears across the conductor according to the relationship:

$$V = BLv \qquad (4.19)$$

BASIC PRINCIPLES OF A DC MACHINE

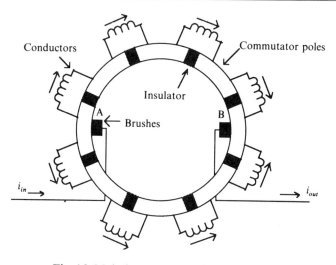

Fig. 4.9 Multple current-carrying conductors.

where:

V = voltage
v = the velocity of the conductor as it moves through the magnetic field.

In a DC motor, the armature has many wraps of current-carrying wires. As illustrated in Fig. 4.9, current is passed in through brushes which contact a series of slip rings, called a commutator. There are many commutators segments, called "poles", which are separated by insulating materials. As the commutator rotates, the direction of current flow in the poles and their conductors will change, but the direction of current flow in the flux field is always the same because current will always flow from A to B.

In a practical DC motor the conductors are tightly wrapped together when the motor is constructed. The area of incidence on a single conductor is computed as the total area of incidence on the armature divided by the number of poles.

The surface area of the conductor in the magnetic field, A, is computed as follows:

$$A = \frac{2\pi r L}{p} \tag{4.20}$$

where:

r = the radius of the armature
L = the length of the armature in the magnetic field
p = the number of poles

If there are conductors wrapped in parallel between poles, then equation 4.19 must be multiplied by the number of parallel wraps, n. Looking at the hypothetical arrangement in Fig. 4.9, there are eight poles with one conductor path each. Using equations 4.5, 4.17, 4.18 and 4.20, we can compute the torque created by a single pole of a DC motor:

$$T = F \cdot r = BLir = \frac{\varphi}{A} Lir \qquad (4.21)$$

$$T = \left[\frac{P\varphi}{2\pi}\right] i \qquad (4.22)$$

Note that the quantity in the brackets is strictly a function of the construction of the motor. Once the motor design is determined, that quantity is a constant. Hence,

$$T = K_T i \qquad (4.23)$$

where:

K_T = the torque constant for the motor.

The torque constant is a parameter of the motor published as part of the motor specification.

4.5 OPERATING CHARACTERISTICS OF DC MOTORS

As previously described, in choosing a DC motor for an application an important criterion is the torque that the motor is capable of delivering at various speeds. At low speeds, during acceleration from rest, the motor must provide enough torque to overcome the inertia of the system, resistance from viscous damping, and any work being done during acceleration. At working velocities, the motor must provide enough torque to overcome viscous damping and work being done at constant velocity, e.g. grinding a metal surface. The important operating characteristic of a DC motor is the torque/speed relationship and it is published by the manufacturer as part of the specification of the motor.

For the purposes of this discussion, we shall focus on the permanent magnet DC motor. An important characteristic of this motor is that it has a linear torque/speed relationship. For that reason, it has turned out to be the motor of choice for many control applications.

From equation 4.23, the fundamental relationship between motor torque and current is $T = K_T i$. At zero speed, when the motor is beginning its rotation, the current developed is proportional to the applied voltage and the resistance in the armature:

$$i = \frac{V}{R_a} \qquad (4.24)$$

where:

V = voltage applied to the motor
R_a = armature resistance

As described earlier, when a conductor is moved through a magnetic field, there is a generator effect, where some voltage is created across the conductor. This voltage has polarity that is opposite to the applied voltage and is known as a back electromotive force, or back EMF. In general:

$$E_B = K_B \omega \tag{4.25}$$

where:

E_B = back EMF (volts)
K_B = back EMF constant, determined by the construction of the motor
ω = armature speed (rad/s)

After rotation has begun, the general expression for armature current, including the effect of back EMF, is:

$$i = \frac{V - E_B}{R_a} = \frac{V - K_B \omega}{R_a}$$

The torque/speed relationship for a DC motor can now be written as:

$$T = K_T i = \frac{K_T(V - K_B \omega)}{R_a} \tag{4.26}$$

Figure 4.10 illustrates the linear torque/speed relationship for the permanent magnet DC motor. The limiting conditions occur at zero velocity and at maximum velocity. At zero velocity, $\omega = 0$, and

$$T = \frac{K_T V}{R_a} \tag{4.27}$$

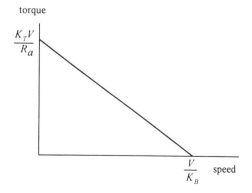

Fig. 4.10 Torque/speed curve of a permanent magnet DC motor.

As back EMF builds up, the motor reaches a stall out condition. At that point, $T = 0$ and $\omega = \omega_{max}$. Therefore,

$$\omega_{max} = \frac{V}{K_B} \qquad (4.28)$$

The operating characteristic, or torque/speed curve, is an important consideration in matching motor performance to loading conditions.

EXAMPLE 4.5

A permanent magnet DC motor has a constant voltage of 24 volts applied. At stall the motor draws 6 amperes; when rotating at 60 rad/s the motor draws 1.05 amp. What is the armature resistance, R_a, and back EMF constant, K_B?

Answer

$$i = \frac{V - E_B}{R_a}$$

At stall, $E_B = 0$, $R_a = 24/6 = 4$ ohms.
 At 60 rad/s,

$$1.05 = \frac{24 - K_B(60)}{4}; \quad K_B = 0.33 \text{ volts/rad s}^{-1}$$

EXAMPLE 4.6

The motor of Example 4.5 has a value $K_T = 0.33$ oz-in/amp. What torque will the motor deliver at 60 rad/s?

Answer

$$T = K_T i = \frac{K_T(V - K_B \omega)}{R_a} = \frac{0.33(24 - 0.33(60))}{4}$$

$$T = 0.3465 \text{ oz-in}$$

EXAMPLE 4.7

A mechanical system has viscous damping of $f = 0.02$ oz-in/rad s^{-1} and a mechanical load with torque $T_L = 0.3$ oz-in. Can the motor of Example 4.6 be used to drive this system at a constant velocity of 20 rad/s when 24 volts are applied?

Answer

Torque (mechanical system) = $T_L + f\omega = 0.3 + 0.02(20) = 0.7$ oz-in

The maximum torque that can be delivered by the motor at 20 rad/s is:

$$\text{Torque (motor)} = \frac{K_T(V - K_B\omega)}{R_a} = \frac{0.33(24 - 0.33(20))}{4} = 1.4355 \text{ oz-in}$$

We conclude that, ignoring inertia, this motor is capable of the application.

4.6 SPEED CONTROL OF DC MOTORS

In this section we describe some basic methods of speed control for DC motors. This will also serve as foundation material for an understanding of control theory, which is the subject of Chapter 5.

Since torque and speed are proportional to the current flowing in the armature of a DC motor, the simplest and most direct method of speed control is to control the magnitude of the current. This can be done by adding a variable resistor in series with the armature, as shown in Fig. 4.11.

The current passing through the armature is now determined by

$$i = \frac{V}{R_a + R_1}$$

where R_1 is determined by the operator's setting on the potentiometer. The resulting family of torque/speed curves is illustrated in Fig. 4.12 and is computed using

$$T = \frac{K_T(V - K_B\omega)}{R_a + R_1} \qquad (4.29)$$

The dotted line of Fig. 4.12 indicates changing speeds for a given (constant) torque requirement as R_1 is increased.

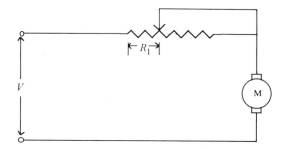

Fig. 4.11 Speed control by varying resistance.

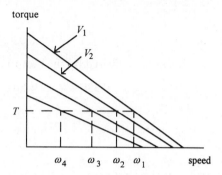

Fig. 4.12 Family of torque/speed curves for various values of R_1.

This method of speed control, though simple and inexpensive, gives relatively poor regulation of speed. As R_1 changes, there is a changing level and slope in the torque/speed curve. In addition, the system is inefficient due to power losses in the potentiometer.

Using a variable resistor to regulate speed does not lend itself to direct computer control. For this you need to regulate the input voltage itself. Figure 4.13a shows an open loop control system for the regulation of motor speed by computer control. In Fig. 4.13a, the computer provides an 8-bit binary output to the D/A converter, which converts the binary count to a voltage level. Here

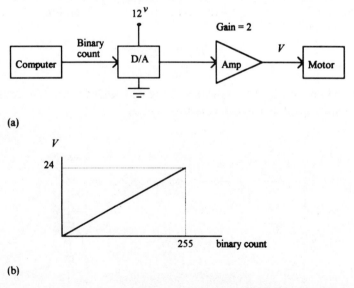

Fig. 4.13 Computer regulation of motor speed by regulating input voltage. (a) Computer control of input voltage; (b) mapping between computer output and motor input.

SPEED CONTROL OF DC MOTORS

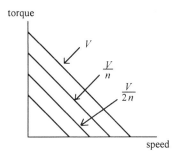

Fig. 4.14 Family of torque/speed curves from regulating voltage, $n > 1$.

we illustrate a full range of 0 to 12 volts. The signal is amplified by twice and applied to the motor. By regulating the number in the output port of the computer, the voltage input to the motor is regulated. Figure 4.13b shows the mapping between the 8-bit binary count of the computer and the input voltage of the motor. However, as we have seen in Chapter 3, the mapping function is not actually linear. In this case there is a step size of 0.094 volts.

The family of torque/speed curves resulting from controlling the input voltage is shown in Fig. 4.14. The slope remains constant while the curve shifts toward the origin as voltage is lowered.

Another approach to speed regulation is pulse width modulation (PWM). PWM is well suited to computer control because it relies on switching a voltage source on and off at high speeds. As illustrated in Fig. 4.15a, a single computer output bit is connected to a switch mode voltage regulator. This device is operated by switching a transistor on and off at high speeds. The result is that the voltage V appears at the output with the same frequency as the computer is providing the input pulse train. The net effect on supply voltage is shown in Fig. 4.15b.

PWM switches the supply voltage on and off. If this is accomplished at high-enough speeds, the motor will not show any discernable changes in speed.

If the frequency of the pulse train is T and voltage is on for T^*, then the average voltage across the circuit is

$$V_{\text{avg}} = V\left(\frac{T^*}{T}\right) \quad (4.30)$$

Consequently, the motor torque is

$$T = K_T I_{\text{avg}} = \frac{K_T(V_{\text{avg}} - K_B \omega)}{R_a} \quad (4.31)$$

The ratio $T^* T^{-1}$ is called the "duty cycle" and speed control is achieved by controlling the length of the duty cycle.

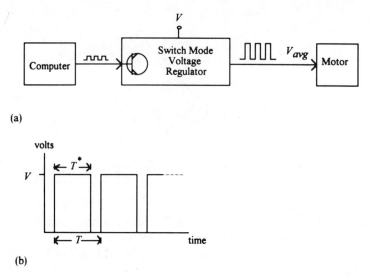

Fig. 4.15 Control of motor speed using pulse width modulation. (a) Computer control with pulse width modulation; (b) PWM duty cycle.

4.7 STEPPING MOTORS

Stepping motors have become important in the era of computers because they are inherently digital devices. They are incremental positioning devices that are moved by rotating a magnetic field around a stator. Figure 4.16 is a simplified illustration of a permanent magnet DC stepping motor. The armature consists of a permanent magnet with north/south poles. The stator consists of a series of electromagnets that can be energized by current flow in the windings. The direction of current flow will determine the polarity of the stator pole.

In Figure 4.16a, the armature is at rest, since the armature and stator are aligned for maximum magnetic attraction. In Fig. 4.16b, polarity is reversed on one stator pole such that a new equilibrium position exists. The forces set up by reversing polarity have caused a rotation of the stator 45 degrees counterclockwise. The value of 45 degrees is what is referred to as the **step angle**, the smallest increment of angular motion over which the motor can be controlled.

In Fig. 4.16c and 4.16d, the motor has completed two more steps. This sequence of motion in which one pole is energized at a time is referred to as "half stepping". In Fig. 4.17, another sequence is shown in which opposing poles change simultaneously. The angle of movement is doubled. This sequence is referred to as "full stepping".

STEPPING MOTORS

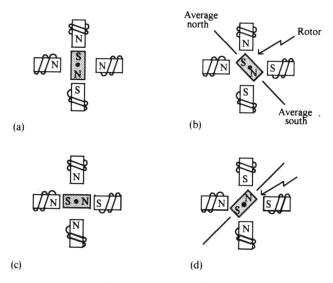

Fig. 4.16 Stepping motor half step sequence.

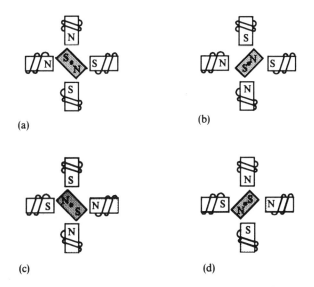

Fig. 4.17 Stepping motor full step sequence.

Controlling the stepping angle depends on the switching sequence applied to the stator coils. Figure 4.18 shows a circuit that can be used to drive the hypothetical motor just described. Coils 1A, 2A, 1B, and 2B are the stator electromagnet coils. The transistors are forward biased when the base is

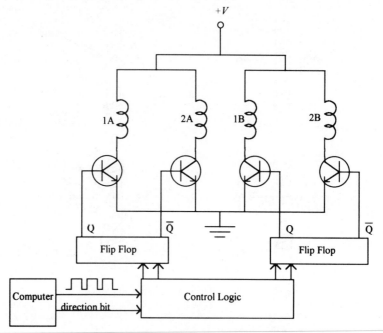

Fig. 4.18 Unipolar drive for stepping motor.

energized. The base is energized by the logic state of the flip flops which are, in turn, controlled by a pulse input through some logic gating. The pulse input is typically provided by a computer or industrial controller. The combination of transistors and logic devices used to switch voltages on and off are referred to as a **translator module**. The computer provides two inputs to the translator module. One input is a control bit (0 or 1) that determines whether the motor rotates clockwise or counterclockwise. The other input is a pulse train that determines the motor speed in steps per second based on the number of pulses per second from the computer.

In Fig. 4.19 the logic states of the flip flops are illustrated for a stepping sequence in the counterclockwise direction. The xs mark which flip flop outputs are energizing transistors. In the drive circuit of Fig. 4.18 only the stator poles whose coils are energized will have polarity. The remaining poles are de-energized. For that reason, this is called a unipolar drive circuit. In Figs 4.16 and 4.17 we illustrated the case where the north/south conditions of a pole could be switched and all poles were energized at each step. This requires a bipolar drive.

In a commercial stepping motor unit, a motor and a drive are specified. Interfacing to a computer or other digital controller is very straightforward. Typically this is accomplished by direct wiring of two or three output bits from

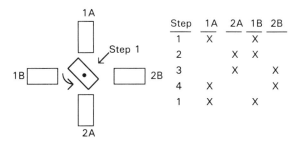

Fig. 4.19 Stepping sequence for drive of Fig. 4.18.

the computer. The direction of rotation is chosen by putting a logic high (or low) on the direction bit, indicating clockwise or counterclockwise rotation. The state of the flip flops is changed by providing a pulse train over the drive bit. The frequency of the pulse train, which is controlled by user programming in computer software, will determine the speed of the motor. Note that the pattern of Fig. 4.19 used to illustrate a counterclockwise rotation. When the controller indicates clockwise rotation, the pattern is simply run in reverse.

Commercial stepping motors require fairly precise angular positioning. Therefore, the number of stator poles would have to be significantly increased over our hypothetical example. Hence, stepping motors are designed with a large number of stator and rotor teeth. Typical stepping motors have step angles that range from 0.72 degrees to 15 degrees. The larger step angle motors are used in such devices as printers, while high-resolution, smaller step angle motors are required in laboratory measurement devices and some machine tool applications. Table 4.1 lists several popular step angle configurations.

As in the case of DC motors, stepping motors are specified in terms of their torque/speed characteristics. Step motor speed is usually specified in terms of steps per second as opposed to revolutions per minute. Since machine design

Table 4.1 Typical step angles for stepping motors

Step angle (degrees)	Steps per revolution
15.0	24
5.0	72
2.5	144
2.0	180
1.8	200
0.72	500

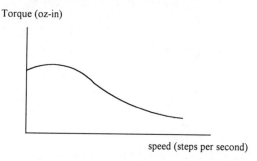

Fig. 4.20 Torque/speed curve for stepping motor.

speeds are usually thought of in terms of rpm, the following conversion is necessary:

$$\text{Steps per revolution} = \frac{360°}{\text{step size}}$$

$$\text{rpm} = \frac{\text{steps per second}}{\text{steps per revolution}} \times 60 \text{ s/min}$$

In Fig. 4.20, a typical torque/speed curve is shown for a stepping motor. As opposed to the permanent magnet DC motor, this torque/speed curve is nonlinear.

Stepping motors have the benefit of being useful in a wide range of open loop operations. Since the position of the motor shaft is determined by a series of pulses, position can be accurately specified and kept track of at the computer. However, if the motor is employed at high speeds or with high inertial loading, there exists the possibility that the motor may overshoot its intended position due to momentum present at stopping. In that case positional accuracy is lost and any further movement begins from an undetermined point.

Stepping motors can be used in closed loop control to improve accuracy where there is fear of losing steps. A typical implementation is to put an incremental encoder on the motor shaft. The incremental encoder puts out a pulse with each step of the motor. These input pulses to the computer are used to control the output pulse train from the computer to the motor. The pulse train is not allowed to continue until an input pulse is received from the encoder.

4.8 RELAY SWITCHES AND SOLENOIDS

Among the simplest of electric actuators are relay switches and solenoids. The operation of a relay switch is illustrated in Fig. 4.21. A set of contacts is wired

RELAY SWITCHES AND SOLENOIDS

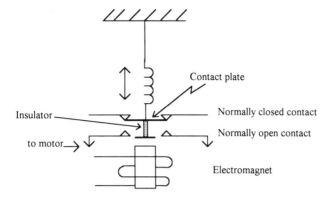

Fig. 4.21 Simplified illustration of a relay.

to a circuit that is to be controlled in an on/off fashion. For example, the circuit might be the starting circuit for a motor on a machine tool. When at rest the contact may be either open (normally open) or closed (normally closed). An electromagnet is used to provide the closing force. When energized, the metal plate is attracted downward and the contact is closed, thus completing the circuit and starting the motor.

Practical relay switches usually require voltages above those which can be supplied directly by a computer; therefore, direct control by the output signal of a computer is not possible. Figure 4.22 is a typical design for computer control. The computer is optically isolated from the relay circuit. When the

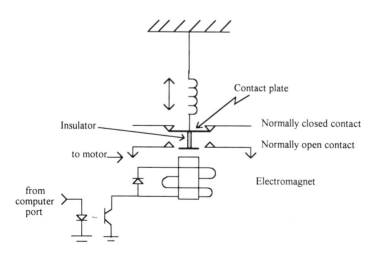

Fig. 4.22 Relay activated by computer.

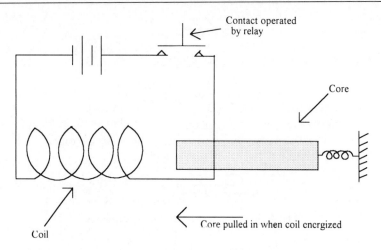

Fig. 4.23 Principle of a solenoid.

computer bit is on, the phototransistor base is biased by the light-emitting diode and current flows through the electromagnet to ground. The computer controls the electromagnet by turning the phototransistor on and off. The diode across the voltage supply prevents voltage spikes from destroying the transistor when the relay opens.

On/off control by relay can have many useful applications. One useful application is in conjunction with a solenoid. An electric solenoid is another electromechanical device that works on the principle of setting up an electromagnetic field. Figure 4.23 illustrates the functioning of a solenoid. The core is normally held part way out of the coil by a spring. When the switch is closed, the coil is energized and the resultant force field pulls the core back into the coil. A typical application of a solenoid might be to control a mechanical valve. When the solenoid retracts, the valve opens.

4.9 FLUID ACTUATORS

Fluid power actuators are very widely used in providing work in mechanical systems and in process control systems. The two most widely used fluid drive systems are hydraulic systems and pneumatic systems. In this section we discuss some common fluid actuators.

4.9.1 Pneumatic cylinders

Pneumatic cylinders exert a linear force through the application of air pressure. Figure 4.24 shows a typical double-acting pneumatic cylinder. The

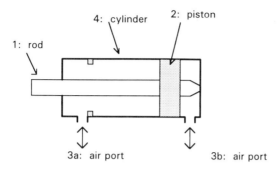

Fig. 4.24 Double-acting pneumatic cylinder.

piston rod provides the external force to perform the work. The rod is moved by the application of air to either side of the piston. When air is applied at port 3a while it is allowed to exhaust at port 3b, the piston rod is driven to the right in the position shown in Fig. 4.24. When air is applied at port 3b while it is allowed to exhaust at 3a, the piston and piston rod are driven to the left. The result is a force applied across a stroke of the cylinder.

The main considerations in choosing a pneumatic cylinder for an application are the required force, stroke and speed. The force developed by the piston can be computed as

$$F = P \cdot A \tag{4.32}$$

where:

F = the force in lbs
P = the air pressure in lb/in^2
A = the effective area of the piston

The **cylinder stroke** is the difference between the fully extended and fully retracted rod arm. Cylinders come in standard stroke sizes, which are described in manufacturers' catalogs.

Cylinder speed depends on the applied load, the cylinder diameter and the rate of air inflow. The rate of air inflow depends on the inlet valve size and the applied air pressure. Pressure regulators are placed in the air line to adjust the rate of air inflow and attain the desired cylinder speed.

Pneumatic cylinders are appropriate for applications in which position control is not required. A pneumatic cylinder has basically two positions, fully extended and fully withdrawn. Intermediate positions are possible using mechanical stops, but control of position throughout the stroke is not possible. The main advantage of pneumatic cylinders is that they are very inexpensive and easy to control.

4.9.2 Solenoid valves

The control of air to a pneumatic cylinder is accomplished through position control of a valve. The valve may be controlled by a motor, but a simple control system requiring only on/off control may be accomplished using a solenoid. The latter is illustrated by Fig. 4.25, which shows a pneumatic control valve that would be appropriate to the operation of the pneumatic cylinder of Fig. 4.24.

In Fig. 4.25a, the solenoid is at its fully retracted position. The pistons are directing the inflow of air such that it would enter air port 3a of Fig. 4.24 and exit air port 3b. When the solenoid is energized and becomes fully extended, air enters air port 3b and exits air port 3a. Thus, the position of the pneumatic

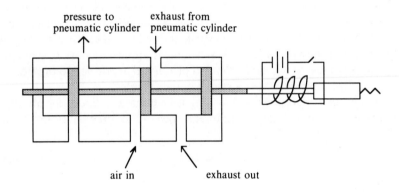

(a)

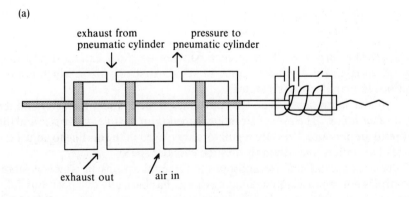

(b)

Fig. 4.25 Two-position, four-way solenoid valve. (a) Shuttle to rightmost position; (b) shuttle to leftmost position.

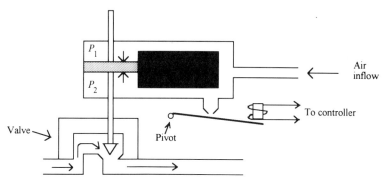

Fig. 4.26 Principle of a flapper valve.

cylinder is controlled electronically through the operation of the solenoid valve.

4.9.3 Flapper valves

A class of valves that allow continuous control of position using differential air pressure is known as a flapper valve and is illustrated in Fig. 4.26. The valve is electronically operated. The position of the flapper determines the amount of air that will escape in the lower channel. The differential air pressure, $P_1 - P_2$, acts on the piston and determines the position of the valve stem. When the gap between the escape port and the flapper is maximum, the valve is in its fully closed position. The fluid flow in the pipe is off. When the gap is fully closed, maximum flow takes place.

4.10 SUMMARY

In this chapter we have described some typical classes of actuators used in manufacturing. These included DC motors, stepping motors, relay switches, solenoids and fluid actuators. These actuators are used as final control devices in manufacturing processes, such as machines and fluid control systems. When they are combined with transducers that monitor the state of the process, a closed loop control system can be designed. This is the subject of the next chapter.

APPENDIX 4A: Quantities and their units

Symbol	Definition	SI (metric) units	British (imperial) units
f	Viscous damping	Nm/rad s^{-1}	oz-in/rad s^{-1}
F	Force	N	oz, lb
g	Gravitational acceleration	m/s^2	in/s^2, ft/s^2
I	Inertia	kg m^2	oz-in-s^2
K_B	Back EMF constant	volts/rad s^{-1}	volts rad s^{-1}
K_T	Torque constant	Nm/amp	oz-in/amp
m	Mass	kg	$\text{oz-s}^2/\text{in}$
T	Torque	Nm	oz-in, lb-ft

EXERCISES

1. A load is being moved by a rack and pinion as shown in Fig. 4.27. The following data applies:

   - weight of load (including rack) = 5 lb
   - radius of gear = 2 in
   - desired maximum velocity = 4 ft/s
   - time to reach velocity = 1 s
   - pinion inertia = 0.2 oz-in-s^2
   - motor rotor inertia = 0.1 oz-in-s^2
   - frictional forces between surface and weight = 6 oz

   (a) Compute the torque required to overcome inertia.
   (b) Compute the torque required to perform the work at constant velocity.
   (c) What is the minimum torque/speed specification for the motor?

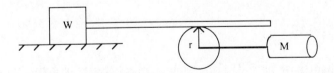

Fig. 4.27 Rack and pinion.

2. Revise the data of Exercise 1 to exclude a maximum velocity and a specified time to reach maximum velocity. Instead, the total length of the move is 10 feet and the load must traverse that distance in 5 seconds. What is the

EXERCISES

minimum torque/speed specification for the required motor assuming a move profile like that of Fig. 4.3?

3. A table is being positioned in the x-axis as shown in Fig. 4.28. The drive motor is a stepping motor wih a step angle of 0.72°. The inertia of the motor rotor and drive system is 300×10^{-3} oz-in-s². The inertia of the load is 1000×10^{-3} oz-in-s². Assume a frictionless system.

   (a) It is desired to bring the table from rest to a speed of 1 ft/s in 0.5 s. What is the required specification of the motor?
   (b) What is the linear resolution of the system?

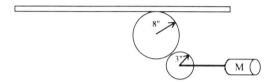

Fig. 4.28 Stepping motor positioning a table.

4. Calculate the torque in oz-in required to accelerate and raise a weight using a drum and pulley (see Fig. 4.29) as follows:

   - weight = 80 oz (5 lb)
   - drum diameter = 3 in
   - drum weight = 4 lb
   - required steady state velocity = 5 ft/s
   - time to reach velocity = 0.5 s
   - motor rotor inertia = 2.5×10^{-3}

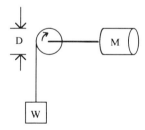

Fig. 4.29 Drum and pulley.

5. Using the data of Exercise 4 and not requiring that maximum velocity be reached in 0.5 s, calculate the motor requirements if the total distance of the move is 50 ft and the time of the move is 8 s.

6. A pulley is being used to move a block 2 ft up an inclined plane, as shown in Fig. 4.30. The block weighs 10 lb and the angle of the incline is 45°. The radius of the drum is 0.5 ft. Assume a frictionless surface.

   (a) How much work is required for this move?
   (b) How much torque must be applied at the drum if the move is to be made in 2 s?

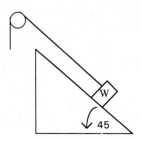

Fig. 4.30 Weight on an incline.

7. We want to design a table that can be positioned by a lead screw driven by a stepper motor. The lead screw pitch is 2 threads per inch and has an efficiency of 50%. Frictional forces between the nut and screw at constant velocity is 4 oz and the table weighs 3 lb.

   (a) If the table must be positioned with a resolution of 0.01 in, what should the step angle of the motor be?
   (b) What are the motor requirements if the maximum acceleration is 1 in/s²?

8. A 24-volt permanent magnet DC motor draws 5 amperes when running at 1000 rpm and 2 amperes when running at 4000 rpm. What is the value of the back EMF constant, K_B, and the armature resistance, R_a?

9. When the motor of Exercise 8 is driving a load with a torque of 2 lb-ft, it runs at a speed of 1000 rpm. What is the torque constant, K_T?

10. I can control the motor of Exercise 9 using pulse width modulation. With 24 volts applied, what duty cycle will allow me to drive the motor at a speed of 1000 rpm?

11. I have a motor with the following parameters:

    $K_B = 1.0\,v/\mathrm{rad\,s}^{-1}$
    $K_T = 3.0\,\mathrm{oz\text{-}in/amp}$
    $V = 24\,\mathrm{volts}$
    $R_a = 4\,\mathrm{ohms}$
    $f = 0.2\,\mathrm{oz\text{-}in/rad}$

I want to drive a load that presents an opposing force of 10 oz and an inertia of 100 oz-in-s². The maximum velocity will be 8 rad/s.

(a) If I drive the load using a rack and pinion with a 6-inch gear, can I achieve my objectives? Explain how.
(b) If I further require the system to reach maximum velocity in 5 s, can I achieve my objectives? Explain how.

12. A permanent magnet DC motor is tested as follows. With a constant voltage of 20 volts applied, the motor draws 5 amps when stalled. When rotating at 50 rad/s, it draws 1 amp. What are the armature resistance and back EMF constant?
13. The motor of Exercise 12 exerts 1 ft-lb of torque when running at 40 rad/s. What is the value of the motor torque constant, K_T?
14. A linear valve can be characterized by the equation $Q(t) = K\theta(t)$, where:

$Q(t)$ = the fluid flow rate at time t in in^3/s
K = a valve constant in $inches^3$/degree-s
$\theta(t)$ = the valve position at time t in degrees

For this problem, $K = 0.4$ and θ has a range of 0–180 degrees. If I control the valve stem using a stepper motor, what is the required step angle to control the flow rate with a resolution of 0.1 in^3/s?
15. For Fig. 4.4, assume that there are four gears between the load and the motor with radius r_1, r_2, r_3, r_4. Derive an expression for the reflected torque of the load in terms of gear ratios.

FURTHER READING

de Silva, C. W. (1989) *Control Sensors and Actuators*, Prentice-Hall, Inc., Englewood Cliffs, New Jersey.
Electro-Craft Corporation (1980) *DC Motors, Speed Controls, Servo Systems*, 5th edn, Electro-Craft Corporation, Hopkins, Minnesota.
Johnson, C. D. (1988) *Process Control Instrumentation Technology*, 3rd edn, John Wiley & Sons, Inc., New York.
Lentz, K. W. (1994) *Design of Automatic Machinery*, 2nd edn, Chapman & Hall, London.
Parker Hannifin Corporation (1991) *Positioning Control Systems and Drives*, Parker Hannifin Corporation, Rohnert Park, California.

5 Control theory

5.1 INTRODUCTION

In Chapter 1 we introduced the concept of a closed loop control system. In a typical closed loop system, knowledge of the current state of the output is used to automatically adjust the control variable so that the output reaches some desired steady state. Examples of closed loop systems occur in the temperature control of a chemical reaction, the speed control of a robot arm, and the position control of the table of a computer-controlled milling machine.

In this chapter we shall introduce the mathematical modeling of closed loop system control. Beginning with the analog system, typical control strategies used in the control of actuators are shown. These are followed by a description of digital control and its relationship to classical analog theory.

5.2 COMPONENTS OF A CONTROL SYSTEM

Figure 5.1 is a block diagram that illustrates the basic components of a feedback control system. There are four components of interest in this system: the actuator and process, feedback loop, summing point and control action. The **actuator** is the system component being directly controlled and which, in turn, controls that aspect of the process we wish to regulate. The action of the actuator on the **process** yields an actual output, O_a, which is the variable of interest to the controller. This may be the actual temperature of a process, actual velocity of the joint of a robot arm, or the actual position of the table of a machine tool. The actual output is compared to a desired output, which is set as O_d by the operator of the process, e.g. a production worker. The comparison of actual output to desired output is done through the feedback loop, typically a transducer monitoring the actual output. The feedback loop of Fig. 5.1 is known as a unity gain feedback loop because the output signal is not altered when it is fed back; i.e. the amplification multiplier is unity. Feedback loops may be designed to provide other than unity amplification of the feedback signal.

The **summing point** is a device that compares the desired output with the actual output and generates an error signal based on the comparison. In Fig. 5.1, a simple difference error is taken, i.e. $e = O_d - O_a$.

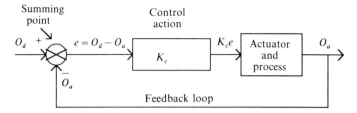

Fig. 5.1 Basic components of a feedback control system.

The error signal is manipulated by a controller and may be boosted through an amplifier before it is applied to the process. The manner in which the error signal is manipulated is called the **control action.** The control action in Fig. 5.1 is simply to take the error signal and multiply (amplify) it by K_c. The manipulated and amplified error signal is then used to control the actuator.

Figure 5.2 illustrates the principles of Fig. 5.1 applied to the position control of a DC motor. The process in this example is the load that is being positioned by the actuator (motor). Hence, the servomotor and load correspond to the block labeled 'actuator and process' in Fig. 5.1.

The actual output is the angle θ_a and the desired output is the angle θ_d. The actual angular position is fed back by a transducer, in this case a rotary potentiometer transducer. The output of the transducer is a voltage, V_a, which is proportional to actual angular position. The 'desired' position of the output shaft can be selected by a worker who turns a rotary dial on a control panel to θ_d. In fact, this rotary dial turns a rotary potentiometer that outputs a voltage, V_d, that is proportional to desired angular position. These two voltages are differenced by the summing point circuit. The summing point produces an output $e = V_d - V_a$, as shown. This voltage is amplified to $K_c e$ and used to drive the motor in the proper direction based on the polarity of the error signal until the desired position is attained, i.e. until $e = 0$. At that point the motor is in a stable, or steady state, position. This simple example is useful in illustrating

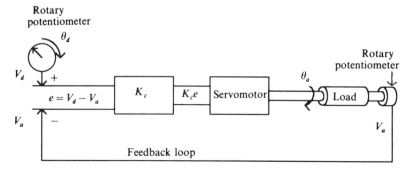

Fig. 5.2 Position control of a DC motor.

the general model applied to a specific system. In the following sections we describe an approach to modeling closed loop control problems in general. We shall use the example of the permanent magnet DC motor extensively to illustrate general principles.

5.3 MATHEMATICAL CHARACTERIZATION AND TRANSFER FUNCTIONS

When analyzing the behavior of any system it is desirable to characterize it mathematically. The mathematical description of the components of a system allows the designer to analyze system performance based on component selection. In control theory the mathematical model of a system component is known as a transfer function. It is a mathematical function that relates the output of the component to its input.

Each block of the block diagram of Fig. 5.1 represents a component for which a transfer function must be developed. Without exception, the block labeled 'actuator and process' is the most difficult component for which to develop a transfer function. This is the case because it requires a detailed understanding of the physics of the combined actuator and process. With each new process, the physics must be defined to model the transfer function. In order to illustrate this we shall develop a transfer function based on the physics of a common actuator presented in the previous chapter, i.e. the description of the relationships that account for the working of a permanent magnet DC motor.

Assume we are interested in the angular velocity, ω, at the output of a DC motor. Recall from the previous chapter that we defined the relationship between torque, acceleration and velocity for a mechanical system driven by a permanent magnet DC motor:

$$T = I\alpha + f\omega + T_L \tag{5.1}$$

where:

T = torque generated at the motor shaft
I = inertia of load and motor, providing mechanical torque proportional to acceleration
α = angular acceleration
f = damping of system, providing mechanical torque proportional to velocity
ω = angular velocity
T_L = torque due to loading, not proportional to acceleration or velocity

It has also been previously shown that the torque produced by a DC motor is proportional to the current flowing in the armature.

$$T = K_T i_a \tag{5.2}$$

where:

K_T = the motor torque constant
i_a = the armature current

When equations 5.1 and 5.2 are combined for a motor with only inertia and damping torques (a motor operating under no load, $T_L = 0$),

$$K_T i_a = I\alpha + f\omega \tag{5.3}$$

The speed of the motor is controlled by controlling the armature current.

As shown in a previous chapter, the speed of a DC motor creates a back EMF, which reduces the current flowing in the armature circuit for a given applied voltage. In particular,

$$E_B = K_B \omega$$

where:

E_B = back EMF voltage generated at velocity ω
K_B = back EMF constant

Figure 5.3 shows the important physical relationships that determine the speed of a DC motor: the mechanical torque from the inertia and damping in the motor, the motor torque generated by the current flowing in the armature of the motor, and the applied voltage and back EMF generating current flow in the armature of the motor. These are time varying relationships, and should be written as follows:

$$T(t) = I\frac{d\omega(t)}{dt} + f\omega(t) \tag{5.4}$$

$$T(t) = K_T i_a(t) \tag{5.5}$$

$$V(t) = Ri_a(t) + E_B(t) \tag{5.6}$$

$$E_B(t) = K_B \omega(t) \tag{5.7}$$

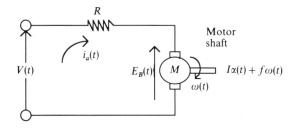

Fig. 5.3 Determinants of the speed of a DC motor.

The velocity transfer function of the motor is a mathematical relationship between the output of the motor, $\omega(t)$, and the input to the motor, $V(t)$, in terms of system design parameters. By substituting equation 5.7 into equation 5.6,

$$V(t) = Ri_a(t) + K_B\omega(t) \tag{5.8}$$

Equating equation 5.4 and equation 5.5.

$$i_a(t) = \frac{I\dfrac{d\omega(t)}{dt} + f\omega(t)}{K_T} \tag{5.9}$$

which, when substituted into equation 5.8 and simplified yields:

$$K_T V(t) = RI\frac{d\omega(t)}{dt} + (Rf + K_T K_B)\omega(t) \tag{5.10}$$

Equation 5.10 is an equation of the time domain response of the speed of a DC motor in relation to the voltage input. To simplify the analysis of equation 5.10 it is desirable to employ **Laplace transforms**, which reduces the analysis of integrodifferential equations to algebra.

5.4 LAPLACE TRANSFORMS

The use of Laplace transforms is somewhat analogous to the use of logarithms for multiplication and division. The product of two numbers can be obtained by adding their logs. This is followed by taking the antilog to obtain the numerical result. Similarly, if an integrodifferential equation written as a function of time is transformed into the s (frequency) domain, where s is the Laplace operator, a solution can be obtained using ordinary algebraic manipulation. The answer can then be transformed back to the time domain.

The Laplace transform, $F(s)$, of $f(t)$ is defined as:

$$F(s) = \int_0^\infty f(t)e^{-st}\,dt$$

Thus, to find the Laplace transform of any function $f(t)$, the function is multiplied by e^{-st} and then integrated from 0 to ∞.

For example, if $f(t) = A$, then:

$$F(s) = \int_0^\infty Ae^{-st}\,dt = A\frac{-1}{s}e^{-st}\bigg|_0^\infty = \frac{A}{s},$$

which is a step function of magnitude A, as shown in Table 5.1.

If $f(t) = \sin \omega t$,

$$F(s) = \int_0^\infty \sin\omega t\, e^{-st}\,dt$$

LAPLACE TRANSFORMS

Table 5.1 Laplace transforms

Time function	Laplace transform
$f(t)$	$F(s)$
1 (unit step)	$\dfrac{1}{s}$
A (step of magnitude A)	$\dfrac{A}{s}$
At (ramp of magnitude A)	$\dfrac{A}{s^2}$
e^{at}	$\dfrac{1}{s-a}$
$\sin \omega t$	$\dfrac{\omega}{s^2+\omega^2}$
$\cos \omega t$	$\dfrac{s}{s^2+\omega^2}$
$f'(t)$	$sF(s) - f(0)$
$f^n(t)$	$s^n F(s) - \sum_{r=0}^{n-1} s^{n-1-r} f(0)$
$\int_0^t f(t)\,dt$	$\dfrac{F(s)}{s}$
te^{at}	$\dfrac{1}{(s+a)^2}$

It can be shown that $e^{j\omega t} = \cos \omega t + j \sin \omega t$ and $e^{-j\omega t} = \cos \omega t - j \sin \omega t$, where $j = \sqrt{-1}$. Therefore,

$$F(s) = \frac{1}{2j}\int_0^\infty (e^{+j\omega t} - e^{-j\omega t})e^{-st}\,dt$$

$$= \frac{1}{2j}\int_0^\infty (e^{-(s-j\omega)t} - e^{-(s+j\omega)t})\,dt$$

$$= \frac{1}{2j}\left[\frac{1}{s-j\omega} - \frac{1}{s+j\omega}\right]$$

$$= \frac{\omega}{s^2+\omega^2}$$

Once a Laplace transform is obtained for a function, there is no reason to derive it again. Table 5.1 is a table of commonly encountered functions and their Laplace transforms.

Laplace transforms can be used to model the time domain response of a system. We can demonstrate the application of Laplace transforms by transforming the time domain equations of the permanent magnet DC motor developed in section 5.3.

To convert equation 5.4 to the s domain, we note that $d\omega(t)/dt = s\omega(s) - \omega(0)$, where $\omega(0)$ is the value of ω at time 0. For simplicity, we will assume an initial condition of $\omega(0) = 0$. Hence, in the s domain, equation 5.4 is written:

$$T(s) = Is\omega(s) + f\omega(s) \tag{5.11}$$

Similarly, equations 5.5, 5.6 and 5.7 are rewritten

$$T(s) = K_T i_a(s) \tag{5.12}$$

$$V(s) = R i_a(s) + E_B(s) \tag{5.13}$$

$$E_B(s) = K_B \omega(s) \tag{5.14}$$

By substituting equation 5.14 into equation 5.13 and using the result to eliminate i_a in equation 5.12, we set the result equal to equation 5.11 as follows:

$$K_T \frac{V(s) - K_B \omega(s)}{R} = [Is + f]\omega(s)$$

Or, rearranging terms:

$$K_T V(s) = [K_T K_B + IRs + fR]\omega(s) \tag{5.15}$$

Note that equation 5.15 is written in terms of the input voltage, $V(s)$, and the output shaft speed, $\omega(s)$, and constants of the system. We can then form the output/input relationship, or velocity transfer function, of the motor as follows:

$$\frac{\omega(s)}{V(s)} = \frac{K_T}{K_T K_B + IRs + fR} \tag{5.16}$$

Equation 5.16, which is the speed transform of the DC motor, gives us a convenient way of modeling the speed response of the motor to a voltage input. Figure 5.4 is a block diagram for the control system of a permanent magnet DC motor illustrating the motor transfer function as part of a control loop.

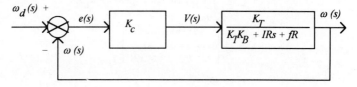

Fig. 5.4 Permanent magnet DC motor in velocity control loop.

USING TRANSFORMS TO ANALYZE SYSTEM RESPONSE 179

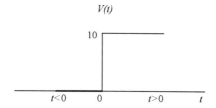

Fig. 5.5 A step input of 10 volts.

5.5 USING TRANSFORMS TO ANALYZE SYSTEM RESPONSE

Using equation 5.16 we can analyze the time domain response of the motor when it is given a voltage input. Assume that input $V(t)$ is a step function of 10 volts as shown in Fig. 5.5. The step function models the instantaneous application of 10 volts to the motor at $t = 0$. The motor will respond to this input by beginning to accelerate. We are interested in knowing the time domain response of $\omega(t)$. From equation 5.16,

$$\omega(s) = \frac{K_T}{K_T K_B + IRs + fR} V(s) \qquad (5.17)$$

and $V(s) = 10\,s^{-1}$ (Table 5.1). Therefore

$$\omega(s) = \frac{10 K_T}{IRs^2 + (fR + K_T K_B)s}$$

Given data concerning the motor, the time domain response can be analyzed.

EXAMPLE 5.1

Assume we have a DC motor with the following parameters (British System of Measurement):

$$K_T = 0.5 \text{ oz-in/amperes}$$
$$K_B = 0.25 \text{ volts/rad s}^{-1}$$
$$R = 4 \text{ ohms}$$
$$I = 0.02 \text{ oz-in-s}^2$$
$$f = 0 \text{ oz-in/rad s}^{-1}$$

Determine the speed response to a step input of 24 volts.

Answer

Note that the motor has no damping forces, $f = 0$. By substitution into equation 5.17:

$$\omega(s) = \frac{0.5}{(0.5)(0.25) + (0.02)(4)s + 0}\left(\frac{24}{s}\right)$$

$$= \frac{150}{s(s + 1.5625)}$$

This can be factored using partial fraction expansion (See Appendix 5A):

$$C_1 = \frac{150}{s(s + 1.5625)} s \bigg|_{s=0} = 96$$

$$C_2 = \frac{150}{s(s + 1.5625)}(s + 1.5625)\bigg|_{s=-1.5625} = -96$$

$$\omega(s) = \frac{96}{s} - \frac{96}{s + 1.5625}$$

Using Table 5.1, a transformation is made back into the time domain:

$$\omega(t) = 96 - 96e^{-1.5625t} \tag{5.18}$$

Figure 5.6 shows the time domain speed response of the motor shaft to a step input of 24 volts under the assumed motor parameters and initial conditions. Such graphs can be used to obtain information about the performance of the motor, such as the time to reach a prescribed velocity.

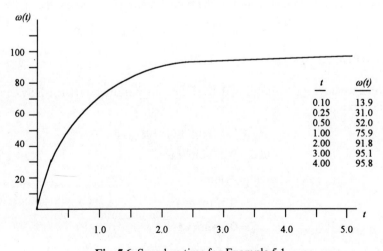

t	$\omega(t)$
0.10	13.9
0.25	31.0
0.50	52.0
1.00	75.9
2.00	91.8
3.00	95.1
4.00	95.8

Fig. 5.6 Speed vs time for Example 5.1.

USING TRANSFORMS TO ANALYZE SYSTEM RESPONSE

Equation 5.18 is a typical response function for a first order process. A first order process is described by the model:

$$\tau \frac{dy}{dt} + y = Kx$$

where:

x = the process input
y = the process output
K = the steady state process gain
τ = the time constant of the process

From equation 5.17,

$$\frac{IR}{K_T K_B + fR} s\omega(s) + \omega(s) = \frac{K_T}{K_T K_B + fR} V(s)$$

$$0.64 s\omega(s) + \omega(s) = 4V(s)$$

Therefore, $\tau = 0.64$ in Example 5.1. Note that the exponent of equation 5.18, 1.5625, is $1/\tau$.

The time constant describes the lag in the process as it approaches steady state. In time τ the process has reached 63.2% of final value. In 3τ and 4τ the process has reached 95% and 98% of final value, respectively. Hence, it takes $3(0.64) = 1.92$ seconds for the process described by equation 5.18 to reach 95% of final value.

The steady state process gain, $K = 4$ in Example 5.1. The transform steady state process gain, when multiplied by the input, yields the steady state output. Note that the steady state output of equation 5.15 is $(4)(24) = 96$.

5.5.1 Relationship between system response and the torque/speed curve of a DC motor

It is instructive to examine Example 5.1 in relation to the material presented in Chapter 4. In Chapter 4 we developed the torque/speed curve of a DC motor from the relationship:

$$T = \frac{K_T(V - K_B \omega)}{R}, \qquad (5.19)$$

where V is the applied voltage. When the motor is unloaded and there are no damping forces ($T = 0$), $\omega_{max} = V K_B^{-1}$. In Example 5.1, we assumed the motor is unloaded at steady state ($f = 0$). Hence, the maximum velocity:

$$\omega_{max} = \frac{V}{K_B} = \frac{24}{0.25} = 96 \text{ rad/s}.$$

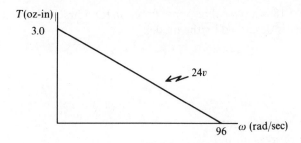

Fig. 5.7 Torque/speed curve for motor of Example 5.1.

Indeed, this is the velocity achieved at steady state as shown in Fig. 5.6 and equation 5.18. We complete the torque/speed curve at $V = 24$ volts by solving for the condition at $\omega = 0$ i.e.:

$$T_{max} = \frac{K_T V}{R} = \frac{(0.5)(24)}{4.0} = 3 \text{ oz-in}$$

Figure 5.7 shows the resulting torque/speed curve.

The duality between the torque/speed curve of the motor and the motor transfer function can now be explained. Equation 5.19, the torque/speed equation, is a steady state equation of the motor for different applied voltages and loads. Thus, $\omega_{max} = 96$ rad/s describes the maximum velocity at no load and an applied voltage of 24 volts.

Equation 5.17, the transfer function, computes the velocity/time path, indicating all the intermediate points in attaining 96 rad/s. Thus, from equation 5.18, at 1.0 s,

$$\omega = 96 - 96e^{-1.5625(1.0)} = 75.9 \text{ rad/s}$$

Using equation 5.19, we can determine the torque delivered by the motor at the speed of 75.9 rad/s.

$$T = \frac{(0.5)[24 - (0.25)(75.9)]}{4} = 0.63 \text{ oz-in}$$

In effect, the motor begins acceleration delivering a torque of 3.0 oz-in. At 75.9 rad/s it is delivering a torque of 0.63 oz-in. When it reaches terminal velocity of 96 rad/s, the delivered torque is zero. This explains the relationship between the torque/speed curve of the motor, which is a static equation, and the transfer function of the motor, which is a dynamic equation.

5.5.2 Motor response with damping

Recall that the motor will not operate above its torque/speed curve. In choosing a motor for a specific steady state speed, it cannot be loaded beyond

USING TRANSFORMS TO ANALYZE SYSTEM RESPONSE

the design load for the specific speed. The speed response function of the motor is also affected by the load condition. This is described in the following example.

EXAMPLE 5.2

In Example 5.1 we assumed that there were no damping forces opposing the motion of the motor. Using the data of Example 5.1, assume the presence of damping forces $f = 0.02$ oz-in/rad s^{-1}. What is the maximum speed attained by the motor when it is subject to a 24-volt step input? How much torque is being delivered at the maximum speed? What is the time domain speed response?

Answer

The motor will accelerate until it reaches a steady state velocity. At that point it will have overcome inertia and will be producing a torque equivalent to the mechanical torque it is opposing. The mechanical torque arises from the damping forces. The equilibrium speed can be computed from the torque/speed equation.

$$T = f\omega = \frac{K_T(V - K_B\omega)}{R}$$

$$(0.02)\omega = \frac{0.5(24 - 0.25\omega)}{4}$$

$$\omega = 58.5 \text{ rad/s}.$$

At the steady state speed of 58.5 rad/s, the torque being delivered is $(0.02)(58.5) = 1.17$ oz-in. To solve for the time domain response, we use the transfer function:

$$\omega(s) = \frac{K_T}{K_T K_B + IRs + fR} V(s)$$

$$\omega(s) = \frac{0.5}{(0.5)(0.25) + (0.02)(4)s + (0.02)(4)} \frac{24}{s}$$

$$\omega(s) = \frac{150}{s(s + 2.5625)}$$

$$\omega(s) = \frac{58.5}{s} - \frac{58.5}{s + 2.5625}$$

$$\omega(t) = 58.5 - 58.5e^{-2.5625t}$$

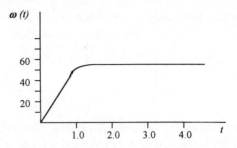

Fig. 5.8 Speed vs time for Example 5.2.

At steady state, the time domain equation gives us the same result as the torque/speed curve. The damping forces of $0.02\,\text{oz-in/rad}\,\text{s}^{-1}$ will limit the motor speed to 58.5 rad/s when 24 volts are applied. It will take approximately 1.56 s for the motor to reach 98% of this speed and, at terminal velocity, it will be delivering $f\omega = (0.02)(58.5) = 1.17$ oz-in of torque. The dynamic behavior of the system is shown in Fig. 5.8.

5.5.3 Motor response under work load

In Example 5.2 we treated the case in which damping forces proportional to velocity opposed motion. There remains the case in which the motor is doing work opposing forces that are not a function of acceleration or velocity, e.g. lifting a load. In Fig. 5.9 we update Fig. 5.3 to include the effect of loading. The load corresponds to the 'process' being controlled by the 'actuator'. For such a case, the inertia plus the torque of the combined actuator and process could be described:

$$T = I\frac{d\omega}{dt} + f\omega + T_L \qquad (5.20)$$

where T_L is the work load. We will assume damping forces are not present ($f = 0$) and equate the mechanical torque (5.20) to the torque delivered by the

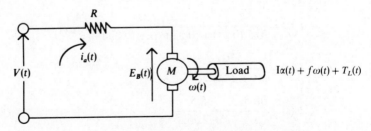

Fig. 5.9 DC motor drive system and load.

USING TRANSFORMS TO ANALYZE SYSTEM RESPONSE

motor (5.19),

$$\frac{K_T(V - K_B\omega)}{R} = I\frac{d\omega}{dt} + T_L,$$

$$\frac{V}{K_B} - \omega = \frac{IR}{K_TK_B}\frac{d\omega}{dt} + \frac{R}{K_TK_B}T_L.$$

Taking the Laplace transform:

$$\frac{V(s)}{K_B} - \omega(s) = \frac{IR}{K_TK_B}s\omega(s) + \frac{R}{K_TK_B}T_L(s)$$

$$\omega(s) = \frac{\dfrac{V(s)}{K_B} - \dfrac{R}{K_TK_B}T_L(s)}{1 + \dfrac{IR}{K_TK_B}s} \qquad (5.21)$$

Equation 5.21 can be used to evaluate the time domain behavior of $\omega(s)$. This equation could have been derived in a manner similar to section 5.4 by including the term $T_L(s)$ in equation 5.11. This derivation will be left as an exercise for the reader.

EXAMPLE 5.3

Assume the motor of Example 5.1 is being used to raise a 0.5 oz weight over a drum at a radius of 2 in from the motor shaft. Assume the drum has no inertia. What is the time response and terminal velocity of the motor? What is the steady state torque?

Answer

The torque and inertia of the weight can be computed:

$$T_L = F \cdot r = (0.5)(2) = 1 \text{ oz-in}$$

$$I\omega = m_wr^2 = \frac{0.5 \text{ oz}}{386.4 \text{ in/s}^2}(2 \text{ in})^2$$

$$= 0.005176 \text{ oz-in-s}^2$$

The total inertia in the system includes the inertia of the motor, which was given in Example 5.1.

$$I = I_m + I_w = 0.02 + 0.005176$$

$$= 0.025176$$

Substituting into equation 5.21:

$$\omega(s) = \frac{\left(\dfrac{1}{0.25}\right)\left(\dfrac{24}{s}\right) - \dfrac{4}{(0.5)(0.25)}\left(\dfrac{1}{s}\right)}{1 + \dfrac{(0.025176)(4)s}{(0.5)(0.25)}},$$

which reduces to:

$$\omega(s) = \frac{79.44}{s(s + 1.24)}.$$

When converted to the time domain,

$$\omega(t) = 64 - 64e^{-1.24t} \qquad (5.22)$$

Equation 5.22 gives the time response of the motor. The terminal velocity is 64 rad/s. We can compute the torque being delivered at 64 rad/s from equation 5.19:

$$T = \frac{0.5[24 - (0.25)(64)]}{4} = 1 \text{ oz-in},$$

which is the torque being required by the mechanical load at steady state, after inertia has been overcome.

5.6 CLOSED LOOP SPEED CONTROL

Referring again to Figure 5.4, the discussion thus far has focused only on the transfer function of the actuator, or motor, as given in the upper right hand block. Equations 5.17 and 5.21 only consider the open loop speed response of the motor to an input voltage. In this section we shall close the loop by developing the transfer function that relates the actual speed response of the motor to a desired speed set by an operator input.

If we refer to Fig. 5.4, we see the operator setpoint, $\omega_d(s)$, is the command signal to the motor, which indicates the desired velocity. This can be an operator setting at the control panel. The equations of the system are written by transforming inputs to outputs around the loop as follows:

$$e(s) = \omega_d(s) - \omega(s)$$

$$V(s) = K_c e(s) = K_c[\omega_d(s) - \omega(s)]$$

$$\omega(s) = \frac{K_T}{K_T K_B + IRs + fR} V(s)$$

$$= \frac{K_T K_c}{K_T K_B + IRs + fR}[\omega_d(s) - \omega(s)]$$

CLOSED LOOP SPEED CONTROL

By substitution and elimination we conclude:

$$\frac{\omega(s)}{\omega_d(s)} = \frac{K_T K_c}{K_T K_B + IRs + fR + K_T K_c} \quad (5.23)$$

Equation 5.23 is the closed loop speed control transform of a permanent magnet DC motor when considering inertia and damping only. It shows the relationship between the actual speed of the motor, $\omega(s)$, to the desired speed setting, $\omega_d(s)$.

EXAMPLE 5.4

Using the data of Example 5.2, assume $K_c = 10$. What is the closed loop speed response of the motor to a desired setting of 40 rad/s?

Answer

We will assume that the operator setting occurs instantaneously i.e. it approximates a step function. Therefore, $\omega_d(s) = 40/s$. Substituting into equation 5.23, we obtain the following:

$$\omega(s) = \frac{(0.5)(10)}{(0.5)(0.25) + (0.02)(4)s + (0.02)(4) + (0.5)(10)} \frac{40}{s}$$

$$\omega(s) = \frac{200}{s(0.08s + 5.205)} = \frac{2500}{s(s + 65.0625)}$$

$$C_1 = 38.4; \quad C_2 = -38.4$$

$$\omega(t) = 38.4 - 38.4e^{-65.0625t} \quad (5.24)$$

5.6.1 Steady state error

There are a couple of things to note about Example 5.4. The first is that, although we requested a speed of 40 rad/s, the system response stabilized at 38.4 rad/s. The difference, 1.6 rad/s, is referred to as **steady state error**. Steady state error is characteristic of first order systems, of which the exponential response function of equation 5.24 is a typical example. One way to eliminate a steady state error is to provide some compensation in the operation of the system. For example, the desired speed of 40 rad/s could be achieved if the operator setting is higher than 40 rad/s.

EXAMPLE 5.5

For Example 5.4, what operator setting is required to achieve a speed of 40 rad/s?

Answer

$$\omega(s) = 40 = \frac{(0.5)(10)}{(0.5)(0.25) + (0.02)(4)s + (0.02)(4) + (0.5)(10)} \frac{\omega_d(s)}{s}$$

$$40 = \frac{5.0\omega_d(s)}{s(0.08s + 5.205)} = \frac{62.5\omega_d(s)}{s(s + 65.0625)}$$

$$C_1 = 0.9606\omega_d(s)$$

$$C_2 = -0.9606\omega_d(s)$$

$$40 = 0.9606\omega_d(s) - 0.9606\omega_d(s)e^{-65.0625t}$$

At $t = \infty$ (SS),

$$40 = 0.9606\omega_d(s)$$

$$\therefore \omega_d(s) = 41.64$$

An operator setting of 41.64 will eliminate the steady state error and provide an output of 40 rad/s. For this system, when the desired steady state value is input, a multiplier of $41.64 \div 40 = 1.041$ will be required in order to eliminate the steady state error. Figure 5.10 extends Fig. 5.4 to show the steady state error compensator, K_e.

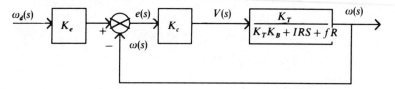

Fig. 5.10 Velocity control loop with steady state error compensation.

EXAMPLE 5.6

Using the data of Example 5.4 and the model of Fig. 5.10 with a value $K_e = 1.041$, what is the speed response to a desired speed of 50 rad/s?

Answer

In this case the error equation follows the relationship

$$e(s) = K_e\omega_d(s) - \omega(s).$$

This results in the closed loop transform:

$$\frac{\omega(s)}{\omega_d(s)} = \frac{K_T K_c K_e}{K_T K_B + IRs + fR + K_T K_c}.$$

CLOSED LOOP SPEED CONTROL

Substituting data:

$$\omega(s) = \frac{(0.5)(10)(1.041)}{(0.5)(0.25)+(0.02)(4)s+(0.02)(4)+(0.5)(10)} \frac{50}{s}$$

$$\omega(s) = \frac{260.25}{s(0.08s+5.205)} = \frac{3253}{s(s+65.0625)}$$

$$\omega(t) = 50 - 50e^{-65.0625t}$$

5.6.2 Mathematical models and physical systems

Another thing to note about Example 5.4, and Example 5.5 as well, is that they are somewhat abstract. That is to say, they do not describe much detail about the physical system that converts the operator input of ω_d to the voltage that actually drives the motor. The physics of the motor has been described at some length, but not the physical devices of the control system. In order to give detailed specification to the mathematical model, we introduce Fig. 5.11.

In Fig. 5.11a, two devices are shown. The operator input is a rotary dial, the shaft of which turns a rotary potentiometer. The readings on the dial, given as ω_d, correspond to a resistance setting on the potentiometer which, in turn, determines the output voltage of the potentiometer. The feedback device is a tachometer, which gives an output voltage proportional to the shaft speed of the motor. The amplifier amplifies the difference between the input voltage, which corresponds to the desired speed set by the operator, and the feedback voltage of the tachometer, which corresponds to the actual speed. The control action amplifies the voltage error by K_c. Hence, $K_c e$ is the voltage used to drive the motor.

In Fig. 5.11b, these components are made explicit in the closed loop control system mathematical model. Note that K_v is a constant that converts the dial setting in rad/s to a voltage. This is designed into the potentiometer dial. Note that K_f is the constant that converts actual speed to volts. This is designed into the tachometer. Figure 5.11 is a specific physical implementation of a closed loop speed controller.

EXAMPLE 5.7

Use the data of example 5.1 and assume that $K_f = 0.3$ volts/rad s^{-1}. What should the value of K_v be in order to reach a desired speed at steady state? Ignore damping forces.

Answer

This question can be answered by evaluating the values of parameters around the control loop of Fig. 5.11b. This is tabulated below. The first column

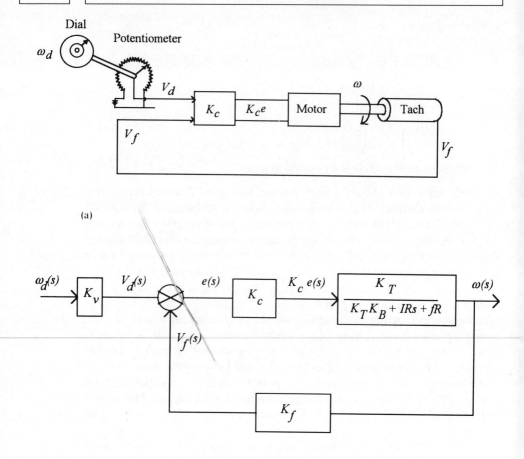

Fig. 5.11 Control system using potentiometer as input and tachometer for feedback. (a) Physical system; (b) mathematical model.

contains various values for the actual speed of the motor, which is determined by $K_c e$, the applied voltage. Since we are ignoring steady state damping forces, equation 5.19 is applied as follows:

$$T = 0 = \frac{0.5(K_c e - 0.25\omega)}{4}$$

$$K_c e = 0.25\omega$$

The values of $K_c e$ are shown in the second column. Since $K_c = 10$, the values of e are computed in the third column.

With $K_f = 0.3$, the feedback voltage can be computed as shown in the fourth column. Since V_d must follow the formula $e = V_d - V_f$, its values are shown in the fifth column.

$\omega(t)$	$K_c e$	e	V_f	V_d
10	2.5	0.25	3	3.25
20	5.0	0.50	6	6.50
30	7.5	0.75	9	9.75
40	10.0	1.00	12	13.00
50	12.5	1.25	15	16.25

Using the above data points, we can solve for K_v. Since, $V_d = K_v \omega_d$, $K_v = 0.325$.

EXAMPLE 5.8

Using the data of Example 5.1 and the model of Fig. 5.11b, what is the speed response of the motor to a setpoint of 75 rad/s?

Answer

From Fig. 5.11b, the following mathematical model can be derived:

$$\frac{\omega(s)}{\omega_d(s)} = \frac{K_T K_c K_V}{K_T K_B + IRs + fR + K_T K_c K_f}$$

$$\therefore \omega(s) = \frac{(0.5)(10)(0.325)}{(0.5)(0.25) + (0.02)(4)s + (0.5)(10)(0.3)} \frac{75}{s}$$

$$\omega(s) = \frac{121.875}{s(0.08s + 1.625)} = \frac{1523.4375}{s(s + 20.3125)}$$

$$\omega(t) = 75 - 75e^{-20.3125t}$$

5.6.3 The advantage of closed loop control

It is useful to elaborate on what is being achieved in a closed loop system. The closing of the loop is ensuring a certain amount of control over in the response of the system to varying conditions, e.g. varying work loads. This can be illustrated by referring to Examples 5.1 and 5.3. In Example 5.1 we derived the open loop speed response of the motor without work load. In Example 5.3 we added a work load to the same motor. The change reduced the steady state speed from 96 rad/s to 64 rad/s, a reduction of a third. In mechanical design, it is not desirable to allow the loading of the motor to determine the speed to such an extent. We can compare this to a closed loop system. In Example 5.8, the motor without load responded to a setpoint of 75 rad/s by achieving it precisely. From equation 5.21, the motor with load follows the relationship:

$$\left[\frac{IR}{K_T K_B}s + 1\right]\omega(s) = \frac{V(s)}{K_B} - \frac{R}{K_T K_B}T_L(s)$$

Since $V(s) = K_c[K_v \omega_d(s) - K_f \omega(s)]$,

$$\left[\frac{IR}{K_T K_B}s + 1\right]\omega(s) = \frac{K_c K_v w_d(s)}{K_B} - \frac{K_c K_f \omega(s)}{K_B} - \frac{R}{K_T K_B}T_L(s)$$

$$\omega(s)\left[1 + \frac{IR}{K_T K_B}s + \frac{K_c K_f}{K_B}\right] = \frac{K_c K_v \omega_d(s)}{K_B} - \frac{R}{K_T K_B}T_L(s)$$

$$\omega(s) = \frac{K_T K_c K_v \omega_d(s) - R T_L(s)}{K_T K_B + IRs + K_T K_c K_f} \tag{5.25}$$

EXAMPLE 5.9

Consider the data of Example 5.3 and assume $K_c = 10$, $K_v = 0.325$ and $K_f = 0.3$. What is the closed loop speed response of the motor to a step input of 75 rad/s?

Answer

Substituting into equation 5.23, we obtain the following:

$$\omega(s) = \frac{(0.5)(10)(0.325)\left(\frac{75}{s}\right) - 4\left(\frac{1}{s}\right)}{(0.5)(0.25) + (0.025\,176)(4)s + (0.5)(10)(0.3)}$$

$$= \frac{117.875}{s(0.100704s + 1.625)}$$

$$= \frac{1170.5096}{s(s + 16.1364)}$$

$$= \frac{72.54}{s} - \frac{72.54}{s + 16.1364}$$

$$\omega(t) = 72.54 - 72.54 e^{-16.1316t}$$

With a load applied, the motor slows down. However, the speed reduction is not as severe as in the open loop case in which the application of the load reduced maximum speed from 96 rad/s to 64 rad/s, or a reduction of 33%.

Here the speed reduction was 3%. A certain amount of stability in maintaining target values is achieved in a closed loop system. In fact, by judiciously choosing design parameters and control strategy, deviations from target values can be held to a minimum.

5.6.4 Adjustment of motor voltage through feedback

Some insight into how the adjustment process takes place in Example 5.9 can be gained by examining the voltage being applied to the motor, V, in both the open loop and closed loop configuration. The open loop case is shown in Fig. 5.12a. The no load speed is 96 rad/s, which drops to 64 rad/s when a mechanical torque of 1 oz-in is applied.

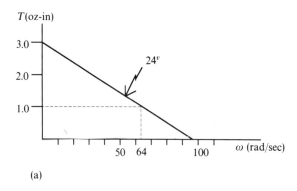

(a)

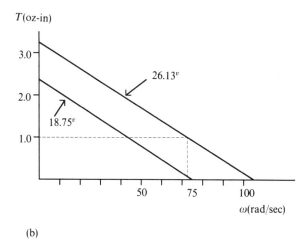

(b)

Fig. 5.12 Torque/speed curves under no load and $T = 1.0$ oz-in. (a) Open loop; (b) closed loop.

For the feedback case, we note that

$$V = K_c e = K_c(K_v \omega_d - K_f \omega)$$

When $\omega_d = 75$ rad/s and there is no load,

$$V = 10[(0.325)(75) - (0.3)(75)] = 18.75 \text{ volts}.$$

When $\omega_d = 75$ rad/s and the load is 1 oz-in,

$$V = 10[(0.325)(75) - (0.3)(72.54)] = 26.13 \text{ volts}.$$

Figure 5.12b shows the torque speed curve for the closed loop case. There are actually two curves. When the motor is unloaded, it is operating on the 18.75 volt curve. When the load is added, it begins to slow down. However, this reduces the feedback voltage from the tachometer, thus increasing the error. The increased error, when amplified, significantly raises the voltage at the motor, which increases the motor speed. The process finally settles at 26.13 volts. The net result of the adjustment process is only a minor reduction in speed. Here it is assumed that the motor can be operated at voltages of 26 volts and higher.

5.6.5 Design considerations

It is also insightful to examine equation 5.25 and example 5.8, which is the closed loop equation without a load. Equation 5.25 includes the load. They only differ by one term, $R\,T_L(s)$. If the design parameters of the system are chosen such that $K_T K_c \omega_d(s) \gg R T_L(s)$, then the response of the system with load will not be significantly different than the unloaded system.

5.7 POSITION TRANSFORM OF A DC MOTOR

A DC servomotor is an important actuator in many automation applications, e.g. robots and computer-controlled machine tools. A servomotor is a motor, the shaft position of which can be controlled, or 'servoed', using a feedback control loop. In this section we shall illustrate the application of the principles in section 5.5 to the analysis of the position response of a DC motor.

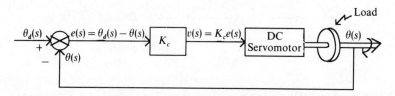

Fig. 5.13 Positioning of a DC motor shaft.

POSITION TRANSFORM OF A DC MOTOR

Figure 5.13 is the general block diagram of the control system components. The motor transfer block transforms the amplified error input, $V(s)$, to the output position $\theta(s)$. Here we will focus only on the motor transfer function, i.e. the block labelled 'DC servomotor'. The transform for this block can be derived in a manner analogous to section 5.4. Rewriting equations 5.4, 5.5, 5.6 and 5.7:

$$T(t) = I\ddot{\theta}(t) + f\dot{\theta}(t) \tag{5.26}$$

$$T(t) = K_T i_a(t) \tag{5.27}$$

$$V(t) = Ri_a(t) + E_B(t) \tag{5.28}$$

$$E_B(t) = K_B \dot{\theta}(t) \tag{5.29}$$

where $\dot{\theta}$ and $\ddot{\theta}$ are the first and second derivatives of position. The corresponding Laplace transforms are:

$$T(s) = Is^2\theta(s) + fs\theta(s) \tag{5.30}$$

$$T(s) = K_T i_a(s) \tag{5.31}$$

$$V(s) = Ri_a(s) + E_B(s) \tag{5.32}$$

$$E_B(s) = K_B s\theta(s) \tag{5.33}$$

By substitution and elimination of terms, the positional transfer function of a DC motor is as follows:

$$\frac{\theta(s)}{V(s)} = \frac{K_T}{K_T K_B s + IRs^2 + fRs} \tag{5.34}$$

Equation 5.34 could have been obtained directly from equation 5.16 by noting that position is the time integral of velocity. From Table 5.1, the integral of $f(t)$ corresponds to multiplying $F(s)$ by $1/s$ (assuming zero initial condition). Hence, multiplying equation 5.16 by $1/s$ yields equation 5.34. Figure 5.14a is

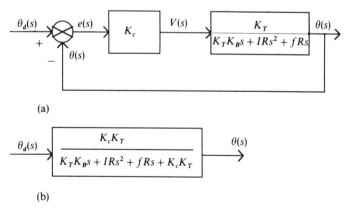

Fig. 5.14 Position control of a DC motor. (a) Closed loop position control of DC motor; (b) transfer function for (a).

a block diagram of the closed loop system, showing the motor transform block.

5.8 POSITION CONTROL OF A DC MOTOR

When a closed loop position control system is combined with the motor transfer function, we can derive the transfer function of the entire system, i.e. $\theta(s)[\theta_d(s)]^{-1}$. From equation 5.34:

$$\frac{\theta(s)}{V(s)} = \frac{K_T}{K_T K_B s + I R s^2 + f R s},$$

and from Fig. 5.14a,

$$V(s) = K_c e(s) = K_c[\theta_d(s) - \theta(s)]. \qquad (5.35)$$

By substituting equation 5.35 into equation 5.34 and reducing,

$$\frac{\theta(s)}{\theta_d(s)} = \frac{K_c K_T}{K_T K_B s + I R s^2 + f R s + K_c K_T} \qquad (5.36)$$

Equation 5.36 is the closed loop transfer function for a DC servomotor with unity feedback. This transfer function is a single representation of the entire system as shown in Fig. 5.14b. It can be used to analyze the time varying behavior of the system.

EXAMPLE 5.10

Assume a DC motor with the following parameters (SI System of Measurement):

$K_T = 8.48 \times 10^{-2}$ Nm/amperes
$K_B = 8.48 \times 10^{-2}$ volts/rad s^{-1}
$R = 0.75$ ohms
$I = 1.696 \times 10^{-5}$ kg m^2
$f = 0$

Using the motor parameters and a feedforward gain, $K_c = 2$, determine the motor response to a unit step input.

Answer

By substitution into equation 5.36 we obtain:

$$\frac{\theta(s)}{\theta_d(s)} = \frac{16.96 \times 10^{-2}}{71.9104 \times 10^{-4} s + 1.2675 \times 10^{-5} s^2 + 16.96 \times 10^{-2}}$$

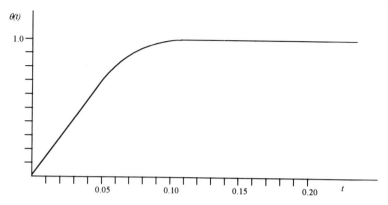

Fig. 5.15 Position vs time for Example 5.10.

For the unit step input, $\theta_d(s) = 1/s$:

$$\theta(s) = \frac{13\,381}{s(s^2 + 567.34s + 13\,381)},$$

which reduces to

$$\theta(s) = \frac{13\,381}{s(s + 24.66)(s + 542.68)}.$$

After a partial fraction expansion,

$$\theta(s) = \frac{1}{s} - \frac{1.047}{s + 24.66} + \frac{0.047}{s + 542.68}.$$

The time domain response is:

$$\theta(t) = 1 - 1.047e^{-24.66t} + 0.047e^{-542.68t}$$

Figure 5.15 is a graph of the response over time of the output shaft position to a unit step input and an initial position of 0. The response achieves the desired angular position of 1 radian.

5.9 CHARACTERISTIC RESPONSES OF A CLOSED LOOP CONTROL SYSTEM

Analysis of the input/output model of a control system can give us much critical information about how the actual system will respond. There are two periods of interest to examine: the transient period and the steady state.

5.9.1 Transient period analysis

When the desired output is set as the input to the closed loop system, there will be a transient period during which the system will try to achieve the desired output. During this period we are interested in the time response of the system. The measures of time response are time delay, rise time and settling time. The **time delay** is the time required for the response to reach 50% of its final value. The **rise time** is the time required for the response to rise from 10% to 90% of its final value. The **settling time** is the time required for the system to settle to within a specified tolerance of its final value.

EXAMPLE 5.11

For the data of Example 5.10 compute the time delay, rise time and settling time to within 1% or final value.

Answer

Time delay:

$$\theta(t) = 0.5 = 1 - 1.047e^{-24.66t} + 0.047e^{-542.68t}$$

$$t = 0.03 \text{ s}$$

Rise time:

$$\theta(t) = 0.9 = 1 - 1.047e^{-24.66t} + 0.047e^{-542.68t}$$

$$t = 0.095$$

$$\theta(t) = 0.1 = 1 - 1.047e^{-24.66t} + 0.047e^{-542.68t}$$

$$t = 0.006$$

$$\text{Rise time} = 0.095 - 0.006 = 0.089$$

Settling time:

$$\theta(t) = 0.99 = 1 - 1.047e^{-24.66t} + 0.047e^{-542.68t}$$

$$t = 0.20$$

Another characteristic of interest during the transient period is the degree of damping that takes place in the system as it tends toward steady state. There are basically four general conditions:

1. no damping
2. underdamped system
3. critically damped system
4. overdamped system.

CHARACTERISTIC RESPONSES OF A CLOSED LOOP SYSTEM

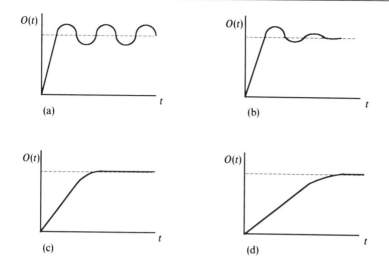

Fig. 5.16 Response functions for different degrees of damping. (a) No damping; (b) underdamped system; (c) critically damped system; (d) overdamped system.

Figure 5.16 illustrates the response for each type of system to a unit step input. An **undamped** system will oscillate continuously. An **underdamped** system will overshoot its steady state position and oscillate with exponential decay until it settles at its steady state value. A **critically damped** system responds in such a way that it reaches its steady state in a relatively short period of time. Like the critically damped system, an **overdamped** system does not oscillate but takes longer to reach its steady state than a critically damped system.

The type of damping displayed by a closed loop control system can be predicted by evaluating the denominator of its transfer function. The denominator of the transfer function is referred to as its characteristic equation and is very important in system analysis. From equation 5.36, the characteristic equation of the DC motor position control is

$$IRs^2 + (K_T K_B + fR)s + K_c K_T \qquad (5.37)$$

or

$$s^2 + \left(\frac{K_T K_B + fR}{IR}\right)s + \frac{K_c K_T}{IR} \qquad (5.38)$$

There are two parameters that are widely used to analyze control systems. They are:

$$\zeta = \text{damping ratio}$$

$$\omega_n = \text{undamped natural frequency}$$

A second order differential equation can be written in terms of these two quantities as follows:

$$f(t) = \frac{d^2y}{dt^2} + 2\zeta\omega_n \frac{dy}{dt} + \omega_n^2 y \qquad (5.39)$$

Comparing equation 5.39 to the characteristic equation for DC motor position control (equation 5.38) implies:

$$\omega_n = \sqrt{\frac{K_c K_T}{IR}} \qquad (5.40)$$

$$\zeta = \frac{1}{2\omega_n}\left[\frac{K_T K_B + fR}{IR}\right] = \frac{K_T K_B + fR}{2\sqrt{K_c K_T IR}} \qquad (5.41)$$

We can give meaning to the two parameters ω_n and ζ. Damping exists in a system when there are forces that tend to oppose motion. If a system characterized by a second order differential equation does not have any forces opposing motion, it will tend to oscillate at its natural frequency as given by ω_n. If the system has forces opposing motion, e.g. viscous damping (f), the system will eventually reach a steady state.

For convenience we will write equation 5.38 as

$$s^2 + 2\zeta\omega_n s + \omega_n^2. \qquad (5.42)$$

When damping is present, the three kinds of damped response can be determined by the following conditions:

$\zeta\omega_n > \omega_n$; $\zeta > 1$ (overdamped)
$\zeta\omega_n = \omega_n$; $\zeta = 1$ (critically damped)
$\zeta\omega_n < \omega_n$; $\zeta < 1$ (underdamped)

From equation 5.41 it can be seen that ζ is a ratio. The denominator of ζ is a measure of critical damping; the numerator of ζ is the amount of damping provided by the damping forces of the system, e.g. back EMF (K_B) and viscous damping (f). If the system's damping exceeds critical damping ($\zeta > 1$) it is overdamped; below critical damping ($\zeta < 1$), it is underdamped. In Example 5.10 the DC motor control system was overdamped. This can be seen in the characteristic response as shown in Fig. 5.15.

The characteristic equation of a system in the form of a second order differential equation is also referred to as a second order process and can be described in terms of its time constants. For $\zeta \leq 1$:

$$\frac{d^2y}{dt^2} + 2\tau^{-1}\frac{dy}{dt} + \omega_n^2 y = \omega_n^2 x$$

where τ = the time constant of the second order process. From equation 5.39 it

CHARACTERISTIC RESPONSES OF A CLOSED LOOP SYSTEM

is clear that
$$\tau = 1\,(\zeta\omega_n)^{-1}$$

EXAMPLE 5.12

Using the transfer function of Example 5.10, illustrate the response to a unit step input if the system does not have any forces opposing motion; i.e. $K_B = f = 0$.

Answer

From Example 5.10, the transfer function with $K_B = f = 0$ is:
$$\frac{\theta(s)}{\theta_d(s)} = \frac{16.96 \times 10^{-2}}{1.2675 \times 10^{-5} s^2 + 16.96 \times 10^{-2}}.$$

When $\theta_d(s) = 1/s$,
$$\theta(s) = \frac{16.96 \times 10^{-2}}{s(1.2675 \times 10^{-5} s^2 + 16.96 \times 10^{-2})}$$

or,
$$\theta(s) = \frac{13\,381}{s(s^2 + 13\,381)}.$$

Factoring the expression $s^2 + 13\,381$ results in imaginary roots and yields a function of the form (see Appendix 5B):
$$\theta(t) = 1 - \cos 115.68t$$

The undamped system starts at an initial value of 0 and oscillates at a natural frequency of $\omega_n = 115.68\ \text{rad/s}$.

EXAMPLE 5.13

For Example 5.10, assume that $K_B = 1 \times 10^{-2}$ and determine the response to a unit step input.

Answer

The proper substitutions result in the equation
$$\frac{\theta(s)}{\theta_d(s)} = \frac{16.96 \times 10^{-2}}{1.2675 \times 10^{-5} s^2 + 8.48 \times 10^{-4} s + 16.96 \times 10^{-2}}$$

From the characteristic equation it is clear that this is an underdamped system since (equation 5.41):

$$8.48 \times 10^{-4} < \sqrt{4(1.2675 \times 10^{-5})(16.96 \times 10^{-2})}$$

Applying a unit step input and rearranging terms:

$$\theta(s) = \frac{13\,381}{s(s^2 + 66.9s + 13\,381)}$$

The above expression is a standard form (see Table 5B.1) where:

$$\omega_n^2 = 13\,381; \quad \omega_n = 115.68$$
$$2\zeta\omega_n = 66.9$$
$$\zeta = 0.289$$

The inverse Laplace transform (Table 5B.1) is:

$$\theta(t) = 1 - 1.045e^{-33.4t}\sin[111t - \tan^{-1}(3.31)]$$

The underdamped system is a fairly typical output for a servo system. Because of inertia in a system, the output cannot both respond quickly to an input command and, at the same time, not overshoot the desired final position. If the system contains too much damping, the rise time may be too long. If the damping is too small, the rise time may be faster but there may be considerable overshoot and a prolonged settling time. System designers must trade off between these and other variables when designing a system for a specific application.

5.9.2 Steady state period analysis

One of the first concerns of system designers is that the control system tends toward a steady state condition. The term 'system stability' is used to characterize the degree to which the transient response will decay over time. The faster this decay occurs under all conditions of operation the system must face (e.g. loading), the more stable the system.

Under some conditions the design of the feedback loop may result in error that drives the system toward further instability. Such unstable systems are avoided in the system design analysis phase through several methods of stability analysis. A discussion of these methods is beyond the scope of this book. The interested reader can refer to the bibliography at the end of this chapter.

Assuming we are working with a stable system, the principal characteristic of interest of the system in the steady state is its final achieved value. This can be determined directly from the Laplace function of the system output using a well-known theorem in control theory called the final value theorem.

CHARACTERISTIC RESPONSES OF A CLOSED LOOP SYSTEM

Final value theorem: The steady state value of $f(t)$ is given by

$$\lim_{t \to \infty} f(t) = \lim_{s \to 0} sF(s)$$

EXAMPLE 5.14

Determine the final achieved value of $\theta(t)$ for the position control problem of Example 5.10.

Answer

Applying the final value theorem:

$$\lim_{s \to 0} s\theta(s) = \frac{13\,381s}{s(s^2 + 567.34s + 13\,381)} = 1$$

When a unit step input is applied, the control system will eventually achieve the desired output.

5.9.3 Analysis of control systems using block diagrams

In section 5.8 and Fig. 5.14 we reduced a control system block diagram to a single transfer function, which we could then easily analyze. In many cases control system block diagrams may be complex, containing several forward and feedback transform blocks. There are several simple rules for combining Laplace transfer functions that may be employed to simplify complex diagrams into single transfer functions. These rules are now described.

(a) Serial blocks

Serial blocks are combined by taking the product of transfer functions and their inputs. The components G_1 and G_2 are Laplace transfer functions that convert inputs to outputs. An equivalent transfer function requires the following operations:

$$W(s) = X(s)G_1(s)$$

$$Y(s) = W(s)G_2(s) = X(s)G_1(s)G_2(s)$$

$$\therefore \frac{Y(s)}{X(s)} = G_1(s)G_2(s)$$

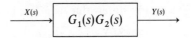

(b) Parallel blocks

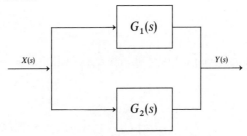

Parallel blocks follow the rules of addition.

$$Y(s) = X(s)[G_1(s) + G_2(s)]$$

$$\frac{Y(s)}{X(s)} = G_1(s) + G_2(s)$$

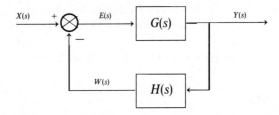

(c) Feedback loops

In feedback loops, summing occurs at the input and the output is being forced to be equal to the input (desired output). When the summing is done using $+$ and $-$ values as indicated on the diagram, the following relations hold:

$$E(s) = X(s) - W(s)$$
$$Y(s) = E(s)G(s)$$
$$W(s) = Y(s)H(s)$$
$$\frac{Y(s)}{G(s)} = X(s) - Y(s)H(s)$$

$$\frac{Y(s)}{X(s)} = \frac{G(s)}{1 + G(s)H(s)} \tag{5.43}$$

CONTROL STRATEGIES

$$\xrightarrow{X(s)} \boxed{\frac{G(s)}{1+G(s)H(s)}} \xrightarrow{Y(s)}$$

EXAMPLE 5.15

Apply the rules of transfer block reduction to Fig. 5.14a to obtain Fig. 5.14b.

Answer

Figure 5.14a has two serial blocks in the feedforward loop and a feedback loop with unit gain. The transfer functions are as follows:

$$G_1(s) = K_c$$

$$G_2(s) = \frac{K_T}{K_T K_B s + IRs^2 + fRs}$$

$$H(s) = 1$$

$$\therefore G(s) = G_1(s)G_2(s) = \frac{K_c K_T}{K_T K_B s + IRs^2 + fRs}$$

and

$$\frac{\theta(s)}{\theta_d(s)} = \frac{G(s)}{1+G(s)H(s)}$$

$$\frac{\theta(s)}{\theta_d(s)} = \frac{K_c K_T}{K_T K_B s + IRs^2 + fRs + K_c K_T}$$

5.10 CONTROL STRATEGIES

In the design of a control system, an important design decision is the kind of control action that will be designed into the controller. In section 5.8 and Fig. 5.14, the control action provided by the controller can be described as simply taking the error between desired and actual output and multiplying it by a constant. This kind of control strategy is known as proportional control, since the control action is linearly proportional to the error term. Proportional control works well in some applications, but can give very unstable responses in other applications. It has been found that three control strategies cover a wide range of engineering applications. They are proportional control, integral control and derivative control. They can also be used in combinations.

Let X_p = the process input, i.e. the output of the control strategy block. Then proportional control is described by

$$X_p(t) = K_p e(t), \tag{5.44}$$

where K_p is the proportional control constant, referred to as the gain.

An integral controller integrates the error signal in computing the process input. The integral controller is described by:

$$X_p(t) = K_i \int e(t)\,dt \tag{5.45}$$

In integral control, the corrective action, X_p, will occur at an increasing magnitude until $e(t) = 0$. This results in a tendency for this kind of control strategy to overshoot the target.

A derivative controller provides a corrective action that is proportional to the rate of change in the error. The derivative control is described by:

$$X_p(t) = K_d \frac{de(t)}{dt} \tag{5.46}$$

For the derivative controller to operate, there must be a changing error signal; if there is constant error the controller will not provide any corrective action. For this reason, derivative control is normally used in conjunction with proportional control or proportional plus integral control.

Proportional-Integral-Derivative (PID) control is a very popular control strategy that combines the three control strategies just described. The PID controller is described by:

$$X_p(t) = K_p e(t) + K_i \int e(t)\,dt + K_d \frac{de(t)}{dt} \tag{5.47}$$

Using Table 5.1 and assuming zero initial conditions, equation 5.47 gives the following Laplace transform:

$$\frac{X_p(s)}{e(s)} = K_p + \frac{K_i}{s} + K_d s \tag{5.48}$$

Figure 5.17a shows the proportional, integral and derivative controllers in a control loop. Combining parallel blocks, Fig. 5.17b shows the conventional representation of a PID controller.

Many commercially available single loop controllers are designed with algorithms for PID control applications. The application engineer using these controllers defines the setpoint of the process and the PID constants. The control equation of these controllers is usually defined as:

$$K_c = K_p \left(1 + T_d s + \frac{1}{T_i s}\right),$$

CONTROL STRATEGIES

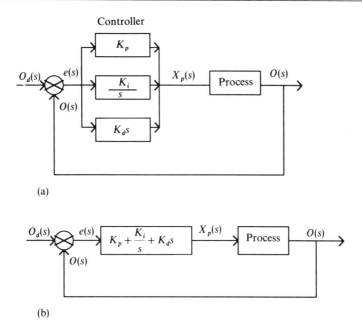

Fig. 5.17 PID control. (a) PID control functions; (b) combined PID control block.

where:

K_p = the proportional gain
T_d = the derivative time constant = K_d/K_p
T_i = the integral time constant = K_p/K_i.

The engineer determines the control action by setting the values of the time constants. Thus, $T_d = 0$ and $T_i = \infty$ yields a proportional controller. As T_d and T_i take on intermediate values, derivative and integral action become part of the control strategy.

EXAMPLE 5.16

Using the DC motor position control problem of Example 5.10, analyze the response to a unit step input with a controller that uses proportional plus derivative control, with $K_p = 2$ and $K_d = 0.2$. For convenience, the other data is reproduced below:

$$K_T = 8.48 \times 10^{-2}$$
$$K_B = 8.48 \times 10^{-2}$$
$$R = 0.75$$
$$I = 1.696 \times 10^{-5}$$
$$f = 0$$

Answer

$$G_1(s) = K_p + K_d s$$

$$G_2(s) = \frac{K_T}{K_T K_B s + IRs^2 + fRs}$$

$$H(s) = 1$$

$$G(s) = G_1(s)G_2(s) = \frac{K_T K_p + K_T K_d s}{K_T K_B s + IRs^2 + fRs}$$

$$\frac{\theta(s)}{\theta_d(s)} = \frac{G(s)}{1 + G(s)H(s)} = \frac{K_T K_p + K_T K_d s}{IRs^2 + (K_T K_d + K_T K_B + fR)s + K_T K_p}$$

$$\frac{\theta(s)}{\theta_d(s)} = \frac{16.96 \times 10^{-2} + 1.696 \times 10^{-2} s}{241.5 \times 10^{-4} s + 1.27 \times 10^{-5} s^2 + 16.96 \times 10^{-2}}$$

Applying a unit step input:

$$\theta(s) = \frac{13\,354.33 + 1335.43s}{s(s^2 + 1901.2s + 13\,354.33)} = \frac{13\,354.33 + 1335.43s}{s(s + 1894.15)(s + 7.05)}$$

$$\therefore \theta(s) = \frac{1}{s} - \frac{0.7}{s + 1894.15} - \frac{0.3}{s + 7.05}$$

$$\theta(t) = 1 - 0.7 e^{-1894.15 t} - 0.3 e^{-7.05 t}$$

When this result is compared to that of Example 5.10, it can be seen that the effect of adding derivative control is to decrease the response time.

5.11 TYPICAL CLASSES OF CONTROL SYSTEM MODELS

The previous sections of this chapter have discussed control theory in the context of motor control. The DC servomotor is an important actuator in manufacturing applications and it is worthwhile understanding the details of its operation and control. However, the theory of feedback control is perfectly general and can be applied to any system for which the appropriate transfer functions can be derived.

In model building, the first order of business is to define the physics of the process under study. The next step is to construct the appropriate block diagram. Having put the problem in standard form, the analysis can be done just as we have shown for the case of the DC motor. In this section we describe some standard problems to illustrate the procedures involved.

5.11.1 A fluid flow problem

Figure 5.18a shows a typical control problem in the process industries, the control of the height of fluid in a tank being filled. The parameters of the system are defined as follows:

$\theta(t)$ = the angular position of a value (in degrees) that controls the input of liquid to the tank at time t
$q_i(t)$ = the rate of flow of liquid into the tank at time t
$q_o(t)$ = the rate of flow of liquid out of the tank at time t
$h(t)$ = the height of liquid in the tank at time t

The physics of this system is as follows. The output of the tank is restricted to a constant cross-section of pipe. Hence, the output flow rate, q_o, varies with the pressure from the height of liquid in the tank. The height of liquid in the tank is governed by the rate of input and the rate of output. The input rate is governed by the valve position. More formally, the defining equations of the system are:

1. Output flow governed by height of water in the tank:

$$q_o(t) = K_o h(t), \qquad (5.49)$$

where K_o = the outlet flow constant.

2. Input flow governed by valve position:

$$q_i(t) = K_x \theta(t), \qquad (5.50)$$

where K_x = the flow control valve constant.

3. The rate of change of the volume (and height) of fluid in the tank is governed by the two flow rates:

$$A \frac{dh(t)}{dt} = q_i(t) - q_o(t) \qquad (5.51)$$

where A = the surface area of the tank.

The three equations above are sufficient to define the process. Substituting equations 5.49 and 5.50 into equation 5.51:

$$A \frac{dh(t)}{dt} = K_x \theta(t) - K_o h(t). \qquad (5.52)$$

Taking Laplace transforms:

$$Ash(s) = K_x \theta(s) - K_o h(s),$$

which yields:

$$\frac{h(s)}{\theta(s)} = \frac{K_x}{As + K_o} \qquad (5.53)$$

Equation 5.52 is the Laplace transform of the process and is appropriately shown in Fig. 5.18b.

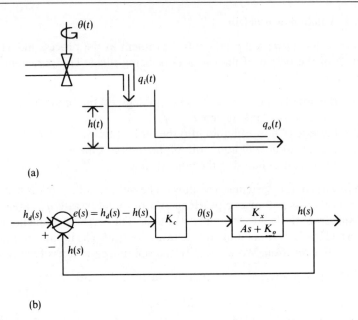

Fig. 5.18 Fluid flow control problem. (a) Physical model; (b) mathematical model.

To control the height of liquid in the tank requires a sensor, e.g. a float, with a servomechanism for adjusting valve position in relation to the desired height of liquid in the tank. For simplicity, assume a unity feedback device is implemented as shown in Fig. 5.18b. With the control model now in standard form the closed loop configuration could be analyzed when appropriate data is given.

EXAMPLE 5.17

For the block diagram of Fig. 5.18, assume the following parameters: $A = 2$, $K_o = 0.2$, $K_x = 0.4$. Compare the response of a proportional controller ($K_p = 2$) with that of an integral controller ($K_i = 2$), when h_d is set to 5.

Answer

For a proportional controller, $\theta(s) = K_p e(s) = K_p[h_d(s) - h(s)]$. Therefore,

$$\frac{h(s)}{h_d(s)} = \frac{K_x K_p}{K_x K_p + As + K_o}. \tag{5.54}$$

TYPICAL CLASSES OF CONTROL SYSTEM MODELS

Substituting data:

$$h(s) = \frac{(0.4)(2)}{(0.4)(2) + 2s + 0.2}\left(\frac{5}{s}\right)$$

$$= \frac{2}{s(s+0.5)}$$

$$h(t) = 4 - 4e^{-0.5t}$$

For an integral controller, $\theta(s) = \frac{K_i}{s}e(s) = \frac{K_i}{s}[h_d(s) - h(s)]$. Therefore,

$$\frac{h(s)}{h_d(s)} = \frac{K_x K_i}{K_x K_i + As^2 + K_o s}$$

$$h(s) = \frac{(0.4)(2)}{(0.4)(2) + 2s^2 + 0.2s}\left(\frac{5}{s}\right)$$

$$= \frac{0.4}{s^2 + 0.1s + 0.4}\left(\frac{5}{s}\right) = 5\left(\frac{0.4}{s(s^2 + 0.1s + 0.4)}\right)$$

Therefore, $\omega_n = 0.6325$ and $\zeta = 0.079$. The time domain response is (Table 5B.1):

$$h(t) = 5\left[1 - \frac{e^{-0.05t}}{0.9969}\sin(0.63t - 1.49)\right],$$

where

$$\tan^{-1}\left[\frac{\sqrt{0.9969}}{-0.079}\right] = 1.49.$$

In comparing the two response functions, note that the proportional controller reaches a steady state value of 4, giving a steady state error of 1. Since the time constant of the system, $\tau = 2$, 98% of final value is reached in 8 seconds. In contrast, the final value of the integral controller is 5. In general, integral controllers eliminate steady state error because, as long as any error exists, the controller continues to sum the error and compensate. However, the relatively low damping coefficient will result in significant overshoot. In addition, the small coefficient on the exponential decay function will result in a relatively long period of oscillation. Since $\tau = 20$, it will take 80 seconds before stabilizing within 2% of final value.

5.11.2 A thermal control problem

A temperature control problem is shown in Fig. 5.19. A controller is adjusting the temperature in an enclosure. A thermocouple feeds actual temperature readings back to the controller.

Assume the equations describing the process have been determined as follows:

$$Q(t) = C\frac{dT}{dt} \tag{5.55}$$

where:

$Q(t)$ = the heat contained in the enclosure at time t
C = the thermal capacity of the enclosure
T = the temperature

The heat flow into the enclosure is determined by the voltage applied to the heating element. Hence:

$$Q_i(t) = K_H V(t) \tag{5.56}$$

where:

$Q_i(t)$ = heat flow into the enclosure
$V(t)$ = the voltage applied to the heating element at time t
K_H = a constant

Heat losses occur through the wall of the enclosure in proportion to the differential temperature across the wall.

$$Q_o(t) = \frac{T(t) - T_a(t)}{R} \tag{5.57}$$

where:

Q_o = heat flow through enclosure surface
T_a = ambient temperature
R = thermal resistance to heat flow through enclosure surface

Finally, the conservation of energy requires that the following equation holds:

$$Q(t) = Q_i(t) - Q_o(t). \tag{5.58}$$

Equations 5.55 to 5.58 are sufficient to characterize the process. By substitution and elimination, the equation relating temperature output to voltage input is

$$T(s) = \frac{RK_H V(s) + T_a(t)}{RCs + 1} \tag{5.59}$$

If we assume a unity feedback loop and a proportional controller:

$$V(s) = K_p[T_d(s) - T(s)] \tag{5.60}$$

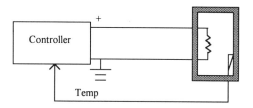

Fig. 5.19 A thermal control problem.

Substituting equation 5.60 into equation 5.59,

$$T(s) = \frac{RK_H K_p[T_d(s) - T(s)] + T_a(s)}{RCs + 1}$$

which reduces to:

$$T(s) = \frac{RK_H K_p T_d(s) + T_a(s)}{RCs + RK_H Kp + 1} \tag{5.61}$$

Given appropriate data, equation 5.61 can be solved for the closed loop performace of the system for a given desired temperature.

The purpose of this section was to illustrate the application of the theory discussed in this chapter to different systems. A number of problems at the end of this chapter provide additional examples for practice.

5.12 DIGITAL CONTROL

A digital control system is one in which at least one element in the control loop is a digital component. Because of the reprogrammability (flexibility) inherent in digital computers, modern industrial controllers are digital systems.

In the previous development of the theory of control, we have assumed a continuous, or analog model. One of the primary differences between an analog and digital system is in the update time of the controller. An analog controller has continuous feedback of the output signal and continuously applies corrective action to the process. A digital system samples the output of the process and applies a control action at discrete points in time. In a digital system, there exists a lag in the process due to the discrete time sampling. This lag adds to the control problem. As the sampling time becomes smaller (approaches zero), the response of the digital system approaches that of the analog system. Hence, when the sampling time is small relative to the time constant of the system, solving the analog model can give a reasonable approximation of the response of the digital system.

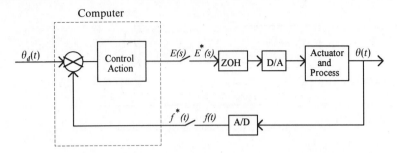

Fig. 5.20 Representation of a digitial control loop.

The method of Laplace transform is an algebra of continuous functions and is not used in the analysis of digital systems. In a later section we will develop the algebra of the z transform, which is analogous to the Laplace transform, but is used in the analysis of discrete systems.

Figure 5.20 is an example of a digital control loop with sampling points indicated by open switches. The digital computer samples the process from the A/D converter by reading the input intermittently. The desired output is specified by an operator set point or by a computer program. The control action is part of the instructions in the computer software. The computer outputs the control signal to a data holding device; in this example a zero order hold (ZOH) device is specified. The zero order hold device holds (latches) the digital output to the D/A converter. This is necessary to continue to drive the process while the computer is taking another sample at the A/D or performing other computations.

The D/A converter provides the continuous analog voltage to the process. Process output is monitored by a sensor and fed to an A/D converter.

There are two features to note about this control loop. The first is that a lag exists between successive updates of the output signal given to the ZOH. The second point is that the ZOH holds a constant level of the control signal over the interval between control actions. The combination of these two factors adds additional requirements for modeling the system over that which exists in an analog system.

In the sections that follow, we shall construct an analytical model for a digital controller. This begins by illustrating the manner in which the sampling process is modeled. We then introduce the z transform for analyzing digital systems. However, before introducing the analytical model, we introduce an intuitive approach to evaluating a digital system based on simulation. This is the subject of the next section.

5.12.1 Simulating a digital system

The effects of sample time lag on system performance can be shown by comparing a continuous system response to a sampled system using a simu-

DIGITAL CONTROL

Table 5.2 A comparison of analog and digital responses using simulation

	Continuous system				Sampled system					
	Open loop		Closed loop		$\Delta t = 1$ second		$\Delta t = 2$ seconds		$\Delta t = 3$ seconds	
t	$h(t)$	$\theta(t)$	$h(t)$	$\theta(t)$	$h(t)$	$\theta(t)$	$h(t)$	$\theta(t)$	$h(t)$	$\theta(t)$
0	0.00	10.00	0.00	10.00	0.00	10.00	0.00	10.00	0.00	10.00
1	1.90	10.00	1.57	6.86	1.90	6.20				
2	3.63	10.00	2.53	4.94	2.90	4.20	3.63	2.74		
3	5.18	10.00	3.11	3.78	3.42	3.16			5.18	0.00
4	6.59	10.00	3.46	3.08	3.70	2.60	3.97	2.06		
5	7.87	10.00	3.67	2.66	3.84	2.32				
6	9.02	10.00	3.80	2.40	3.92	2.16	4.00	2.00	4.05	1.90
7	10.07	10.00	3.88	2.24	3.96	2.08				
8	11.01	10.00	3.93	2.14	3.98	2.04	4.00	2.00		
9	11.87	10.00	3.95	2.09	3.99	2.02			3.985	2.03
10	12.64	10.00	3.97	2.06	3.995	2.01	4.00	2.00		
11	13.34	10.00	3.98	2.04	3.997	2.006				
12	13.98	10.00	3.99	2.02	3.998	2.004	4.00	2.00	3.985	2.03
13	14.55	10.00	3.995	2.01	3.999	2.002				
14	15.07	10.00	4.00	2.00	3.9993	2.0014	4.00	2.00		
15	15.54	10.00	4.00	2.00	4.00	2.00			4.00	2.00

lation. Such a comparison is shown in Table 5.2. Here we use the data for proportional control from Example 5.17, the fluid flow problem.

The first two data sets are based on the continuous model. The first data set is the open loop response function based on equation 5.53 and a fixed valve position of 10 degrees. The second data set is the closed loop response based on equation 5.54 and a desired height of 5 ft. Note that the closed loop controller adjusts the valve position to realize a steady state height of 4 ft.

Thus, the first data set was computed using the transform

$$\frac{h(s)}{\theta(s)} = \frac{K_x}{As + K_o}.$$

$$h(s) = \frac{0.4}{2s + 0.2} \cdot \frac{10}{s} = \frac{2}{s(s + 0.1)}$$

$$h(t) = 20 - 20e^{-0.1t}$$

The second data set was computed from solution to the closed loop system in Example 5.17:

$$h(t) = 4 - 4e^{-0.5t}$$

The last three data sets, which represent a sampled system, are computed for sampling intervals of $\Delta t = 1$ s, 2 s, and 3 s. This is done by computing the values of variables around the control loop at discrete instances of time. From equation 5.52,

$$A\frac{dh(t)}{dt} = K_x\theta(t) - K_o h(t).$$

From Table 5.1, when initial conditions are considered,

$$A[sh(s) - h(0)] = K_x\theta(s) - K_o h(s)$$

$$h(s) = \frac{K_x\theta(s) + Ah(0)}{As + K_o}. \tag{5.62}$$

If the initial condition, $h(0)$, is zero, equation 5.62 will reduce to equation 5.53. We must retain the expression for the initial condition in the simulation model because we will be evaluating the model over successive Δt increments of time. Hence, after $t = 0$, the initial condition will be non-zero for each successive evaluation of the equations.

Without loss of generality, equation 5.62 can be evaluated recursively in increments of Δt using the form:

$$h(s) = \frac{K_x\theta(s) + Ah(t - \Delta t)}{As + K_o}.$$

At $t = \Delta t$, $h(t - \Delta t) = 0$. Applying the data of Example 5.17,

$$h(s) = \frac{0.4\theta(s) + 2h(t - \Delta t)}{2s + 0.2} = \frac{0.2\theta(s)}{s + 0.1} + \frac{h(t - \Delta t)}{s + 0.1}.$$

Since $\theta(s) = \frac{\theta}{s}$,

$$h(s) = \frac{0.2\theta}{s(s + 0.1)} + \frac{h(t - \Delta t)}{s + 0.1}.$$

$$C_1 = \frac{0.2\theta}{s(s + 0.1)} s \bigg|_{s=0} = 2\theta$$

$$C_2 = \frac{0.2\theta}{s(s + 0.1)}(s + 0.1) \bigg|_{s=0.1} = -2\theta$$

$$h(s) = \frac{2\theta}{s} - \frac{2\theta}{s + 0.1} + \frac{h(t - \Delta t)}{s + 0.1}$$

$$h(t) = 2\theta(1 - e^{-0.1\Delta t}) + h(t - \Delta t)e^{-0.1\Delta t} \tag{5.63}$$

Equation 5.63 gives the value of $h(t)$ for a given value of θ held constant over the interval Δt and with an initial value of $h(t - \Delta t)$. From Fig. 5.18, the other equations of the system are:

$$e(t) = h_d(t) - h(t) = 5 - h(t) \tag{5.64}$$

and:
$$\theta(t) = K_p e(t) = 2e(t) = 10 - 2h(t) \qquad (5.65)$$

Beginning with $h(t) = 0$ and recursively solving equations 5.65 and 5.63 yields the values of $h(t)$ for a sampled system with sample time Δt. Consider, for example, the first few entries of Table 5.2 for $\Delta t = 2$ s:

@ $t = 0$, $h(t) = 0$; $\theta(t) = 10 - (2)(0) = 10$.
@ $t = 2$, $h(t) = (2)(10)(1 - e^{-0.2}) + 0e^{-0.2} = 3.63$; $\theta(t) = 10 - (2)(3.63) = 2.74$.
@ $t = 4$, $h(t) = (2)(2.74)(1 - e^{-0.2}) + 3.63e^{-0.2} = 3.97$; $\theta(t) = 10 - (2)(3.97) = 2.06$.

The last three data sets in Table 5.2 illustrate the behavior of the discrete system for increasing values of Δt. The first item to note is the response of each sampled system over the first Δt increment of time. In all cases, the value of $h(t)$ corresponds to that of the open loop continuous system. This is the case because the first value of $\theta(t)$ is 10 degrees in all cases and $\theta(t)$ is held constant over the interval Δt.

The second item to note is that, as Δt becomes larger, the response has a tendency to rise faster and, in the extreme case, will result in overshoot and may yield unstable behavior. For $\Delta t = 3$ s, $h(t) = 5.18$ has exceeded $h_a = 5$ at the end of the first sample interval. This yields the error $e = 5 - 5.18 = -0.18$. Using equation 5.65, $\theta(t) = 2e(t) = -0.36$. Since the valve position is limited to values >0 degrees, the entry in $\theta(t)$ is 0.

The system being simulated here is one with a relatively slow response time. From Example 5.17, the time constant of the system is 2 seconds. When the sample interval is greater that the system time constant, overshoot will result. In general, as the sampling time becomes small in relation to the system time constant, the response of the sampled system will approach that of the continuous system.

The reader should note, from the manner in which the data of Table 5.2 was generated, an assumption that is being made about the manner in which the controller is sampling and updating the process. In particular, look at the calculation made at $t = 2$ seconds. The observed value of $h(t)$ is 3.63. This is the height of the fluid at $t = 2$. With that information, the controller updates the angle of the valve at $t = 2$, i.e. immediately after the sample is taken. In effect, there is an assumption of zero lag between the time the sample is taken and the time the process is updated. The controller repeats that action every 2 seconds. This is a case where the controller is occupied with other matters during the update interval. For example, it may be regulating more than one activity and is programmed to update the valve angle every 2 seconds.

There are other possible operating conditions that can apply in a control problem. For example, the controller may be dedicated to a single process with a small time constant, requiring faster updates. The bulk of the lag may exist between the time the sample is taken and the time the output is applied to the ZOH. For that case, the simulation would require a time shift between the point at which the sample is taken and the point in time when the resulting action updates the process. Using the fluid flow problem with $\Delta t = 2$ as an example, if $h(t)$ were sampled at $t = 2[h(t) = 3.63]$ and Δt were the lag between taking the sample and updating the process, the resulting output $\theta(t) = 2.74$ would be applied at $t = 2 + \Delta t$. By the time the corrective action is applied, the sampled information is out of date. The height of the fluid is higher than 3.63 at $t = 4$. In effect, the controller is applying a corrective action that lags the process variable it is trying to correct. When you combine this with the fact that the corrective action is updated once each Δt, as in Table 5.2, the tendency of the controlled process toward oscillatory behavior is accentuated. In an exercise at the end of the chapter, the reader will be asked to recompute Table 5.2 to consider this interpretation of the controller sampling and process updating activity.

Simulation is a very common approach to analyzing the behavior of sampled data system for various time lags. In the following sections we introduce analytical procedures that can also be used for such analysis.

5.12.2 Representation of a sampling process

The representation of a sampling process is shown in Fig. 5.21. The continuous signal, $f(t)$, is sampled every T time units. The sample is taken over a finite length of time, as indicated in Fig. 5.21b. This is shown as a pulse function. If the duration of the pulse is much shorter than the sampling interval, it can be represented as an impulse function shown in Fig. 5.21c.

The relationship between pulses and impulses can be shown more formally by reference to Fig. 5.22. The single pulse is described by the function:

$$f(t) = \begin{cases} h & 0 \text{ to } t_0 \\ 0 & \text{otherwise} \end{cases}$$

The Laplace transform of a pulse can be obtained by applying the Laplace operator:

$$F(s) = L[f(t)] = \int_0^\infty f(t) e^{-st} dt$$

$$F(s) = \int_0^\infty h e^{-st} dt = h \left[\frac{-e^{-st}}{s} \right]_0^{t_0}$$

$$F(s) = \frac{h}{s}(1 - e^{-t_0 s})$$

A special case of the pulse function is the impulse function, $f*(t)$. It is the limiting case of a pulse as $t_0 \to 0$.

$$L[f*(t)] = \lim_{t \to 0} \left[\frac{h}{s}(1 - e^{-t_0 s}) \right]$$

DIGITAL CONTROL

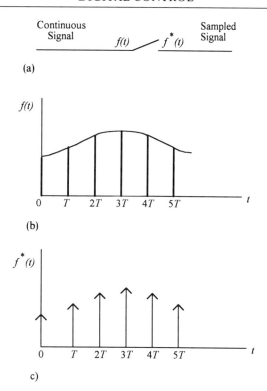

Fig. 5.21 Representation of a sampling process.

For the impulse function of Fig. 5.22, having an area of K, $h = K t_0^{-1}$, and

$$\lim_{t_0 \to 0}\left[\frac{K}{t_0 s}(1 - e^{-t_0 s})\right] = \frac{0}{0},$$

which is indeterminant. An indeterminant function can be evaluated by taking derivatives of the variable in the limit (L'hopital's rule) as follows:

$$L[f^*(t)] = \lim_{t_0 \to 0} \frac{\left(\dfrac{d}{dt_0}\right)[K(1 - e^{-t_0 s})]}{\dfrac{d}{dt_0}(t_0 s)}$$

$$= \lim_{t_0 \to 0} \frac{K s e^{-t_0 s}}{s} = K$$

The Laplace transform of an impulse function is the area of the function. For the unit impulse ($K = 1$), the impulse function is given the special symbol $u_1(t)$. An impulse of area K is written $K u_1(t)$.

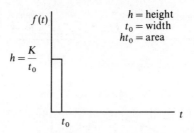

Fig. 5.22 Single pulse.

The mathematical treatment of a digital system usually uses the concept of an 'ideal sampler', which is represented by an impulse stream, $f^*(t)$, of constant time interval, T, as shown in Fig. 5.21c. The area of the impulse at time t, K_t, is equal to the magnitude of the input signal at time t, $f(t)$. Inputs are taken at equally spaced times of duration T. Thus, the area of the nth impulse, which occurs at $t = nT$, is $f(nT)$. Therefore, the entire stream of sampled data is represented by:

$$f^*(t) = f(0)u_1(t) + f(T)u_1(t-T) + f(2T)u_1(t-2T) + \cdots$$

$$f^*(t) = \sum_{n=0}^{\infty} f(nt)u_1(t-nT) \qquad (5.66)$$

where:

$$u_1(t) = \text{unit impulse at } t = 0$$
$$u_1(t-nT) = \text{unit impulse at } t = nT$$

The Laplace transform of equation 5.66 is:

$$F^*(s) = L[f^*(t)].$$

Evaluating the first term in equation 5.66:

$$F^*(s) = L[f(0)u_1(t)] = \lim_{t_0 \to 0}\left[\frac{f(0)}{t_0 s}(1-e^{-t_0 s})\right] = f(0)$$

Evaluating the second term:

$$F^*(s) = L[f(T)u_1(t-T)] = \lim_{t_0 \to 0}\left[\frac{f(T)}{t_0 s}(1-e^{-(t_0-T)s})\right]$$

$$= \lim_{t_0 \to 0} \frac{\dfrac{d}{dt_0}\{f(T)[1-e^{-(t_0-T)s}]\}}{\dfrac{d}{dt_0}(t_0 s)}$$

$$= \lim_{t_0 \to 0} \frac{f(T)se^{-(t_0-T)s}}{s} = f(T)e^{-Ts}$$

DIGITAL CONTROL

In general, the series can be written:
$$F^*(s) = L[f^*(t)] = f(0) + f(T)e^{-Ts} + f(sT)e^{-2Ts} + \cdots$$

The Laplace transform of the ideal samples is

$$F^*(s) = \sum_{n=0}^{\infty} f(nT)e^{-nTs} \qquad (5.67)$$

EXAMPLE 5.18

A continuous signal of the form $f(t) = e^{-at}$ is being sampled every T intervals of time. What is the Laplace transform of the sampled signal, $f^*(t)$?

Answer

Since $f(t)$ is e^{-at}, from equation 5.66, the sampled time domain signal is:

$$f^*(t) = u_1(t) + e^{-at}u_1(t-T) + e^{-2at}u_1(t-2T) + \cdots$$

$$= \sum_{n=0}^{\infty} e^{-ant} u_1(t - nT).$$

From equation 5.67,

$$F^*(s) = L[f^*(t)] = \sum_{n=0}^{\infty} f(nT)e^{-nTs}$$

$$F^*(s) = 1 + e^{-aT}e^{-sT} + e^{-2aT}e^{-2sT} + \cdots$$

$$= \sum_{n=0}^{\infty} e^{-n(s+a)T}$$

5.12.3 Representation of a zero order hold

In section 5.12.2 we examined the model for an ideal sampler. In the context of Fig. 5.20, this was relevant to the sampling process at the A/D converter, i.e. the input side of the controller. In this section we refer to the activities on the output side, in particular, the development of a model for the ZOH function.

From the point of view of the behavior of the process, the important activity is the output control action of the controller and the update time, i.e. the interval of time between successive changes in the controller output. This was the message in Table 5.2, where the successive values of $\theta(t)$, held constant over Δt, determined the state of the process at $t + \Delta t$. An assumption of Table 5.2 is that the update of the output takes place immediately after sampling the input. That is to say, the time lag Δt occurs entirely between the update of θ and taking the next sample. When this simplifying assumption is made, one can assume that $E(s)$ in Fig. 5.20 represents current information on the state of the process at the time the process is updated. This is clearly a simplifying assumption, but one that can be made in developing an approximate model of the discrete controller of the type assumed in Table 5.2, which we do here.

If the system is modeled with $E(s)$ as current information at the time of process update, then the system model is basically the analog model block diagram with an additional block, the zero order hold. If the transform of the ZOH is defined, the system transform can then be determined by combining blocks.

As previously mentioned, a ZOH is a device that retains (latches) the value of the last output until the next output is given. The relationship of the original error signal, sample and ZOH signal is shown in Fig. 5.23.

Figure 5.23a shows the variables involved in the process. $E(t)$ is the actual error between the process setpoint and the actual value of the process at time t. It is a continuous function and is shown in Fig. 5.23b. $E^*(t)$ is the sample of $E(t)$ that is output from the computer. $E^*(t)$ is modeled as a series of impulses. Finally, $h(t)$ is the output of the ZOH. It is of magnitude $E^*(t)$ and exists for duration Δt. The ZOH latches $E^*(t)$ for an interval Δt.

The ZOH converts the time series of impulse functions to a time series of step functions. A step function of magnitude A over the interval 0 to ∞ is notated mathematically as

$$h(t) = Au(t).$$

If the step function exists over the interval 0 to T,

$$h(t) = Au(t) - Au(t-T)$$
$$= A[u(t) - u(t-T)].$$

Therefore, the equation representing a time series of steps corresponding to $h(t)$ is:

$$h(t) = e(0)[u(t) - u(t-T)] + e(T)[u(t-T) - u(t-2T)]$$
$$+ e(2T)[u(t-2T) - u(t-3T)] + \cdots$$

where $e(t)$ is the value of E^* at time t. Since

$$\int_0^T e(0)e^{-st}dt = e(0)\left[\frac{1-e^{-Ts}}{s}\right] \quad \text{and} \quad \int_T^{2T} e(T)e^{-sT}dt = e(T)\left[\frac{e^{-Ts}-e^{-2Ts}}{s}\right],$$

and so on, the Laplace transform is:

$$h(s) = e(0)\frac{1-e^{-Ts}}{s} + e(T)\frac{e^{-Ts}-e^{-2Ts}}{s} + e(2T)\frac{e^{-2Ts}-e^{-3Ts}}{s} + \cdots$$

$$h(s) = \frac{1-e^{Ts}}{s}[e(0) + e(T)e^{-Ts} + e(2T)e^{-2Ts} + \cdots]$$

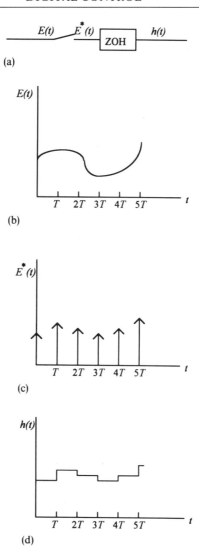

Fig. 5.23 Representation of a ZOH.

The summation within the brackets is analogous to equation 5.67, which is that of the ideal sampler. Therefore,

$$h(s) = \frac{1 - e^{-Ts}}{s} E^*(s) \qquad (5.68)$$

Equation 5.68 combines the ideal sampler, $F^*(s)$, and the ZOH, $1 - e^{-Ts}(s)^{-1}$. Having defined the ZOH transform, we have developed the equation for the

ZOH block of Fig. 5.20. However, evaluating the block diagram for a discrete controller is usually done using the method of the z transform, which will be defined in the next section.

5.12.4 The z transform

As previously stated, a control system with digital components is not analyzed using the Laplace transform. Instead, it is analyzed using the z transform. A sequence $f(nT)$ is transformed to the function $F(z)$ by:

$$F(z) = \sum_{n=0}^{\infty} f(nT) z^{-n} \tag{5.69}$$

EXAMPLE 5.19

Consider the function $f(nT) = a^n$. What is the z transform?

Answer

Since

$$F(z) = \sum_{n=0}^{\infty} f(nT) z^{-n} = \sum_{n=0}^{\infty} a^n z^{-n} = \sum_{n=0}^{\infty} \left(\frac{a}{z}\right)^n$$

$$\sum_{n=0}^{\infty} X^n = \frac{1}{1-X} \quad \text{if } |X| < 1,$$

$$F(z) = \frac{1}{1 - \frac{a}{z}} = \frac{z}{z - a} \quad \text{if } \left|\frac{a}{z}\right| < 1.$$

EXAMPLE 5.20

Consider the unit step function, $f(nT) = 1$ for all n. What is the z transform?

Answer

$$F(z) = \sum_{n=0}^{\infty} f(nT) z^{-n} = \sum_{n=0}^{\infty} z^{-n} = 1 + z^{-1} + z^{-2} + \cdots$$

Since

$$\sum_{n=0}^{\infty} a^n = \frac{1}{1-a}$$

$$F(z) = \frac{1}{1-z} = \frac{z}{z-1}.$$

Table 5.3 z transforms

Time function	Laplace transform	z transform
	e^{st}	z
$u(t)$	$\dfrac{1}{s}$	$\dfrac{z}{z-1}$
t	$\dfrac{1}{s^2}$	$\dfrac{Tz}{(z-1)^2}$
e^{-at}	$\dfrac{1}{s+a}$	$\dfrac{z}{z-e^{-at}}$
$1-e^{-at}$	$\dfrac{a}{s(s+a)}$	$\dfrac{z(1-e^{-aT})}{(z-1)(z-e^{-aT})}$
te^{-at}	$\dfrac{1}{(s+a)^2}$	$\dfrac{Tze^{-aT}}{(z-e^{-aT})^2}$

As in the case of Laplace transforms, there exists a table of the z transforms (Table 5.3) to enable rapid conversion of standard forms.

There are three properties of the z transform that are useful to know; linearity, time shifting, and multiplication by a constant. The linearity property states that any linear combination of time domain functions has a dual linear combination of z domain functions. If

$$F(t) = c_1 f_1(t) + c_2 f_2(t),$$

then

$$F(z) = c_1 F_1(z) + c_2 F_2(z).$$

The time shifting property is one of the most important properties of the z transform. Consider the two time series shown in Fig. 5.24, where 5.24b is simply 5.24a shifted by m units. The relationship is:

$$g(nt) = f(nT - mT) = \begin{cases} f(nT) & \text{if } mT \leqslant nT \leqslant \infty \\ 0 & \text{if } 0 \leqslant nT < m \end{cases}$$

Therefore,

$$F(z-m) = \sum_{n=0}^{\infty} f(n-m)z^{-n} = f(-m) + f(1-m)z^{-1} + f(2-m)z^{-2} + \cdots$$
$$+ f(0)z^{-m} + f(1)z^{-(m+1)} + \cdots$$

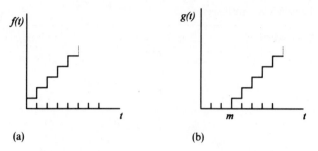

Fig. 5.24 Shifting property of Z transformation.

Since $f(n-m) = 0$ for all $n < m$,

$$F(z-m) = z^{-m}[f(0) + f(1)z^{-1} + f(2)z^{-2} + \cdots]$$

$$\therefore F(z-m) = z^{-m}F(z).$$

Lastly, the z transform of a number sequence multiplied by a constant is equal to the constant multiplied by the z transform of the number sequence.

$$z[af(n)] = az[f(n)] = aF(z).$$

A relationship exists between the z transform and the Laplace transform which can be illustrated by comparing equations 5.69 and 5.67. Any number sequence of the z transform can be written:

$$F(z) = \sum_{n=0}^{\infty} f(nT)z^{-n} = f(0) + f(1)z^{-1} + f(2)z^{-2} + \cdots \quad (5.70)$$

From equation 5.67, the Laplace transform of an ideal sampler provides a number series:

$$F^*(s) = \sum_{n=0}^{\infty} f(nT)e^{-nTs} = f(0) + f(T)e^{-Ts} + f(2T)e^{-2Ts} + \cdots \quad (5.71)$$

The similarity between the two transforms is clear. If we assume the numbers in $F(z)$ were obtained from sampling the function $f(t)$, then $F(z) = F^*(s)$ when $e^{sT} = z$. Given a Laplace transform in e^{sT}, it can be converted to a z transform with the above substitution.

5.12.5 Example using the z transform

For purposes of illustration and comparison with analog control systems, we introduce Fig. 5.25, which is a digital control diagram using the data of Example 5.4. The lag in the sampling of output of the process and subsequent correcting of the error signal at ZOH is incorporated in the switch $E(s)/E^*(s)$.

DIGITAL CONTROL

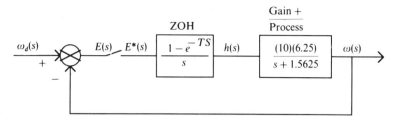

Fig. 5.25 Example of closed loop digital system.

Every T periods of time (the sampling frequency) the controller takes a sample of the process and immediately updates the control signal, which is latched by the ZOH. The critical difference between the digital system of Fig. 5.25 and the analog system of example 5.4 is that $E^*(s)$ will be held constant over each interval nT to $(n+1)T$.

Solving the feedforward transform:

$$G(s) = \frac{\omega(s)}{E^*(s)} = \frac{(10)(6.25)(1-e^{-Ts})}{s^2 + 2.5625s}$$

$$= \frac{62.5(1-e^{-Ts})}{s(s+2.5625)}$$

Expanding the polynomial part using partial fractions:

$$C_1 = \frac{62.5}{s(s+2.5625)} s \bigg|_{s=0} = 24.39$$

$$C_2 = \frac{62.5}{s(s+2.5625)}(s+2.5625) \bigg|_{s=-2.5625} = -24.39$$

$$\therefore G(s) = \frac{24.39(1-e^{-Ts})}{s} - \frac{24.39(1-e^{-Ts})}{s+2.5625}$$

$$= 24.39(1-e^{-Ts})\left(\frac{1}{s} - \frac{1}{s+2.5625}\right).$$

By substituting $z = e^{sT}$ and, from Table 5.2,

$$\frac{1}{s} = \frac{z}{z-1}; \quad \frac{1}{s+a} = \frac{z}{z-e^{-at}}:$$

$$G(z) = z[G(s)] = 24.39(1-z^{-1})\left(\frac{z}{z-1} - \frac{z}{z-e^{-2.5625T}}\right).$$

$$\therefore G(z) = \frac{24.39(1-e^{-2.5625T})}{z-e^{-2.5625T}} = \frac{\omega(z)}{E^*(z)} \quad (5.72)$$

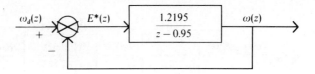

Fig. 5.26 Digital control system with feedforward transform.

For illustrative purposes we will assume that the sampling frequency, $T = 0.02$ seconds. Equation 5.72 reduces to:

$$\frac{\omega(z)}{E^*(z)} = \frac{1.2195}{z - 0.95} \tag{5.73}$$

The transfer function at this stage of reduction is shown in Fig. 5.26. We can include the feedback loop by noting that:

$$E(z) = \omega_d(z) - \omega(z).$$

Substituting into equation 5.73 yields:

$$\omega(z) = \frac{1.2195\omega_d(z)}{z + 0.2695}$$

As in Example 5.4, we apply a step input of 40 rad/s. From Table 5.3 the z transform is:

$$40\left(\frac{z}{z-1}\right)$$

$$\therefore \omega(z) = \frac{48.78z}{z^2 - 0.7305z - 0.2695} \tag{5.74}$$

The time domain response at increments of $T = 0.02$ seconds can be obtained by division of polynomials in equation 5.74. The process is shown below.

$$
\begin{array}{r}
48.78z^{-1} + 35.63z^{-2} + 39.18z^{-3} + 38.22z^{-4} + 38.48z^{-5} + 38.40z^{-6} \\
z^2 - 0.7305z - 0.2695 \overline{\smash{\big)}\, 48.78z} \\
\underline{48.78z - 35.63 \quad -13.15z^{-1}} \\
35.63 \quad +13.15z^{-1} \\
\underline{35.63 \quad -26.03z^{-1} \quad -9.60z^{-2}} \\
39.18z^{-1} \quad +9.60z^{-2} \\
\underline{39.18z^{-1} \quad -28.62z^{-2} \quad -10.56z^{-3}} \\
38.22z^{-2} \quad +10.56z^{-3} \\
\underline{38.22z^{-2} \quad -27.92z^{-3} \quad -10.30z^{-4}} \\
38.48z^{-3} \quad -10.30z^{-4}
\end{array}
$$

The response of the time series in increments of 0.02 s is:

$$\omega(z) = 48.78z^{-1} + 35.63z^{-2} + 39.18z^{-3} + 38.22z^{-4} + 38.48z^{-5} + 38.40z^{-6} + \cdots \tag{5.75}$$

The time series shows underdamped behavior. It overshoots at $T = 0.02$, undershoots at $T = 0.04$, and then begins to settle at the target value. From Example 5.4, the corresponding continuous equation is:

$$\omega(s) = 38.4 - 38.4e^{-65.0625t}$$

For comparison, we juxtapose the analog and digital responses in the following table.

	$\omega(t)$	
t	Analog	Digital
0.02	27.95	48.78
0.04	35.55	35.63
0.06	37.63	39.18
0.08	38.19	38.22
0.10	38.34	38.48
0.12	38.38	38.40

Note that the time lag induced by the sampling process results in a faster rise time for the digital system. As the sampling time is reduced, the digital response will approach that of the analog system. As the sampling time is increased, the digital system will show a greater tendency to overshoot and oscillate. In general, if the sampling interval is small relative to the time constant of the system, solving the analog problem can yield a reasonably good solution for a digital system. This can be followed by fine-tuning the parameters of the system experimentally.

5.13 SUMMARY

In this chapter we have provided an introduction to the subject of control theory. Classical analog theory was described and example problems were solved using the method of Laplace transforms. This was followed by an introduction to digital control, emphasizing the influence of the sampling process on information lag and system stability. This was illustrated using a simulation approach and also using an analytical approach and the z transform method.

CONTROL THEORY

This chapter has merely touched on the subject of closed loop control. Due to the introductory nature of this text, many important issues were left out that the reader may want to pursue later. These include design methods for ensuring system stability, feedforward control and the control of multiple loop systems. The bibliography at the end of this chapter will direct the reader to several texts that provide complete treatment of these subjects.

APPENDIX 5A: PARTIAL FRACTION EXPANSION

In factoring the equation of Example 5.1, we applied the method of partial fraction expansion. In general, a Laplace transform can be expressed as a ratio of two polynomials:

$$F(s) = \frac{b_m s^m + \cdots + b_2 s^2 + b_1 s + b_0}{a_n s^n + \cdots + a_2 s^2 + a_1 s + a_0} \tag{5A.1}$$

Finding the equivalent time domain function requires factoring equation 5A.1. We first factor the denominator:

$$F(s) = \frac{b_m s^m + \cdots + b_2 s^2 + b_1 s + b_0}{(s+r_1)(s+r_2)\cdots(s+r_n)} \tag{5A.2}$$

where $-r_1, -r_2, \ldots, -r_n$ are the roots of the denominator. Equation 5A.2 is then factored into the form:

$$F(s) = \frac{c_1}{s+r_1} + \frac{c_2}{s+r_2} + \cdots + \frac{c_n}{s+r_n} \tag{5A.3}$$

To find $c_1, c_2, \ldots, c_n$, equation 5A.3 is multiplied by the denominator of c_i and s is set equal to the root of the denominator of c_i:

$$c_1 = \frac{b_m s^m + \cdots + b_2 s^2 + b_1 s + b_0}{(s+r_1)(s+r_2)\cdots(s+r_n)}(s+r_1)\bigg|_{s=-r_1}$$

$$c_2 = \frac{b_m s^m + \cdots + b_2 s^2 + b_1 s + b_0}{(s+r_1)(s+r_2)\cdots(s+r_n)}(s+r_2)\bigg|_{s=-r_2}$$

$$\vdots$$

$$c_n = \frac{b_m s^m + \cdots + b_2 s^2 + b_1 s + b_0}{(s+r_1)(s+r_2)\cdots(s+r_n)}(s+r_n)\bigg|_{s=-r_n}$$

Applying this procedure to Example 5.1,

$$\omega(s) = \frac{150}{s(s+1.5625)}$$

$$\omega(s) = \frac{c_1}{s} + \frac{c_2}{s+1.5625}$$

$$c_1 = \frac{150}{s(s+1.5625)} s \bigg|_{s=0} = 96$$

$$c_2 = \frac{150}{s(s+1.5625)}(s+1.5625) \bigg|_{s=-1.5625} = -96$$

In some cases the denominator of a factored term may have squared terms in it. When this occurs, it is necessary to evaluate the squared term twice. Assume, for illustration, the following form:

$$\omega(s) = \frac{150}{s^2(s+1.5625)}$$

It is necessary to factor the squared term twice as follows:

$$\omega(s) = \frac{c_1}{s^2} + \frac{c_2}{s} + \frac{c_3}{(s+1.5625)}$$

The evaluation procedure requires determining c_2 by differentiating the evaluation expression.

$$c_2 = \frac{d}{ds}\left[\frac{150}{s^2(s+1.5625)} s^2\right]\bigg|_{s=0}$$

$$= \frac{d}{ds}\left[\frac{150}{s+1.5625}\right]\bigg|_{s=0}$$

$$= -\frac{150}{(s+1.5625)^2}\bigg|_{s=0} = -61.44$$

$$c_3 = \frac{150}{s^2(s+1.5625)}(s+1.5625)\bigg|_{s=-1.5625} = 61.44$$

$$c_1 = 96$$

APPENDIX 5B: SOLUTION OF A LINEAR DIFFERENTIAL EQUATION WITH IMAGINARY ROOTS

The second order differential equation of a linear position servo system is of the form:

$$\frac{d^2\theta}{dt^2} + 2\zeta\omega_n\frac{d\theta}{dt} + \omega_n^2\theta = 0 \tag{5B.1}$$

The resulting Laplace transform characteristic equation is:

$$s^2 + 2\zeta\omega_n s + \omega_n^2 = 0, \tag{5B.2}$$

The roots of an algebraic equation having the form $ax^2 + bx + c = 0$ are found by the quadratic equation:

$$x = \frac{-b \pm \sqrt{b^2 - 4ac}}{2a} \tag{5B.3}$$

Substituting into equation 5B.3,

$$s = \frac{-2\zeta\omega_n \pm \sqrt{4\zeta^2\omega_n^2 - 4\omega_n^2}}{2},$$

$$s = -\zeta\omega_n \pm \omega_n\sqrt{\zeta^2 - 1}. \tag{5B.4}$$

It is clear from equation 5B.4 that, when the system is overdamped ($\zeta > 1$), s will have real roots. If the system is critically damped ($\zeta = 1$), s will have only one root, $-\zeta\omega_n$. If the system is underdamped ($\zeta < 1$), s will have an imaginary component and equation 5B.4 is written:

$$s = -\zeta\omega_n \pm j\omega_n\sqrt{1 - \zeta^2},$$

where $j = \sqrt{-1}$.

When the roots of the second order system are all real, the response function will be exponential, as in Example 5.10. When imaginary roots exist, as in Examples 5.12 and 5.13, the response will include a sinusoidal function. Table 5B.1, shown below, provides a list of the most common Laplace transforms and their time functions when imaginary roots exist.

Consider the entry for the time domain function $f(t) = 1 - \cos \omega_n t$.

$$F(s) = \int_0^\infty (1 - \cos \omega_n t)e^{-st} dt = \int_0^\infty e^{-st} dt - \int_0^\infty e^{-st} \cos \omega_n t \, dt$$

Using the results from Table 5.1,

$$F(s) = \frac{1}{s} - \frac{s}{s^2 + \omega_n^2} = \frac{\omega_n^2}{s(s^2 + \omega_n^2)},$$

which is the third entry in Table 5B.1, below.

Consider the function given by $f(t) = K^{-1}\omega_n e^{-\zeta\omega_n t} \sin(\omega_n K t)$, where $K = \sqrt{1 - \zeta^2}$. Therefore:

$$f(t) = \frac{\omega_n}{2jK}(e^{-(\zeta\omega_n - jK\omega_n)t} - e^{-(\zeta\omega_n + jK\omega_n)t}),$$

$$L[f(t)] = \int_0^\infty f(t)e^{-st} dt = \frac{\omega_n}{2jK}\int_0^\infty (e^{-(s+\zeta\omega_n - jK\omega_n)t} - e^{-(s+\zeta\omega_n + jK\omega_n)t}) dt,$$

$$L[f(t)] = \frac{\omega_n}{2jK}\left[\frac{1}{s + \zeta\omega_n - jK\omega_n} - \frac{1}{s + \zeta\omega_n + jK\omega_n}\right].$$

EXERCISES

Substituting $\sqrt{1-\zeta^2} = K$ and reducing,

$$L[f(t)] = \frac{\omega_n^2}{s^2 + 2\zeta\omega_n + \omega_n^2},$$

which is the first entry in Table 5B.1.

Table 5B.1 Time domain functions for Laplace transforms having characteristic equations with imaginary roots

Laplace transform	Time function
$\dfrac{\omega_n^2}{s^2+2\zeta\omega_n s + \omega_n^2}$	$\dfrac{\omega_n e^{-\zeta\omega_n t}}{\sqrt{1-\zeta^2}} \sin(\omega_n\sqrt{1-\zeta^2}\,t)$
$\dfrac{\omega_n^2}{s(s^2+2\zeta\omega_n s + \omega_n^2)}$	$1 - \dfrac{e^{-\zeta\omega_n t}}{\sqrt{1-\zeta^2}} \sin[\omega_n\sqrt{1-\zeta^2}\,t - \phi]$, where $\phi = \tan^{-1}[\sqrt{1-\zeta^2}/(-\zeta)]$
$\dfrac{\omega_n^2}{s(s^2+\omega_n^2)}$	$1 - \cos\omega_n t$
$\dfrac{s+a}{(s+a)^2+\omega^2}$	$e^{-at}\cos\omega t$
$\dfrac{\omega}{(s+a)^2+\omega^2}$	$e^{-at}\sin\omega t$

EXERCISES

1. Write the transfer function of the following equations, where x is the input and y is the output.

   (a) $7\dfrac{dy(t)}{dt} = 5x(t)$

   (b) $\dfrac{d^2y(t)}{dt^2} + \dfrac{dy(t)}{dt} + 5y(t) = 2x(t)$

   (c) $te^{-at}y(t) = \dfrac{dx(t)}{dt}$

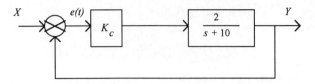

Fig. 5.27 Control loop for Exercise 2.

2. Consider the closed loop control system in Fig. 5.27.
   (a) What is the time domain open loop response if X is a unit step input and K_c is proportional control of $K_p = 10$?
   (b) What is the time domain closed loop response of the system if X is a unit step input?
   (c) What is the time constant, τ, and the process gain, K?
   (d) What is the time delay, rise time, and 1% settling time of part (b)?
   (e) Solve for the time domain closed loop response of the system to an integral control of $K_i = 5.0$.
   (f) What is the steady state value of $e(t)$ for the response of part (e)?
   (g) Solve for the time domain closed loop response of the system to an integral control of $K_i = 50.0$.
   (h) What is the maximum overshoot of the response of part (g)?

3. Expand the following functions of s using partial fraction expansion and solve for $y(t)$.

   (a) $y(s) = \dfrac{4}{2s^2 + 8s + 6}$

   (b) $y(s) = \dfrac{3s + 1}{s(s^2 + 4s + 3)}$

   (c) $y(s) = \dfrac{10\,s^{-1}}{5s^2 + 5s}$

4. A permanent magnet DC motor is operated in a closed loop as shown in Fig. 5.28. If 10 volts is applied at V_{in}, what steady state speed will be achieved? Assume $f = 0.1$, $K_T = 10$, $K_B = 0.2$, $R = 4$, $I = 0.2$.

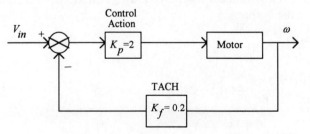

Fig. 5.28 Motor control loop for Exercise 4.

EXERCISES

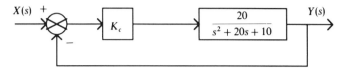

Fig. 5.29 Control loop for Exercise 5.

5. Assume that the block diagram in Fig. 5.29 is a controller for an industrial process, where K_c is the control action.

   (a) Determine the closed loop time domain system response function to a unit step input under proportional control, with $K_p = 2$.
   (b) For part (a) what is the steady state value of $y(t)$?
   (c) Determine the time domain response function to a unit step input under proportional plus derivative control, with $K_p = 2$ and $K_d = 2$.
   (d) For part (c) what is the steady state value of $y(t)$?

6. Shown in Fig. 5.30 is a closed loop velocity controller for a DC motor. The motor is driving a positioning table mounted on a lead screw. The following data applies:

$K_T = 17.6$ oz-in/amperes
$K_B = 0.25$ volts/rad s^{-1}
$R = 4$ ohms
Inertia of motor rotor $= 0.02$ oz-in-s^2
Viscous damping (motor) $= 0.2$ oz-in/rad s^{-1}
Inertia of lead screw and table $= 10$ oz-in-s^2
Gear ratio $= 10:1$
Torque of the load, $T_L = 60$ oz-in
Tachometer constant, $K_f = 0.5$

   (a) Assume an amplifier of gain 100 is used. Derive the transfer function of the motor control loop using symbols (as opposed to numerical data).
   (b) If the gearbox, lead screw, and table were removed from the system, what would be the closed loop motor response to a step input of 24 volts?

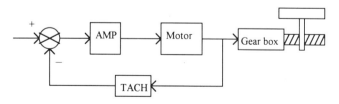

Fig. 5.30 Control problem of Exercise 6.

(c) What is the motor response to a step input of 24 volts with the gearbox, lead screw, and table present in the system?

(d) If the lead screw pitch provides 0.01 in of forward travel per revolution, what is the time response of the table velocity to a step input of 24 volts?

7. The speed control unit of a grinding tool is being studied for speed response characteristics. The tool is driven by a permanent magnet DC motor. When the control is set for 6 rev/s, the data points shown in the following table are collected by a sensor monitoring the speed:

Time (s)	Input speed (rev/s)	Output speed (rev/s)
0.5	6.0	3.33
1.0	6.0	5.65
1.5	6.0	6.09
2.0	6.0	6.05
3.0	6.0	6.00
10.0	6.0	6.00

(a) Characterize the response as either underdamped, critically damped, or overdamped.

(b) The above tests were conducted with inertia and damping present, but no work load (no grinding). It is known that the damping ratio, $\zeta = 0.8$. What is the undamped natural frequency, ω_n? Assume integral control with $K_i = 1$ and unity feedback.

8. For Example 5.17 in the text, compute the value of the steady state compensator required in order that $h(s) = h_d(s)$ when proportional conrol is used.

9. For Example 5.17 in the text, compute the time domain response using a proportional plus derivative control strategy where $K_p = K_i = 2$.

10. Shown in Fig. 5.31 is a system that has been designed to control the rate at which a fluid is pumped. It consists of a DC motor directly driving an impeller. The controller monitors the flow rate at the output of the system using a flow meter. The controller regulates the speed of the motor by the voltage it applies. Assume, for simplicity, that the dynamic equations of the system are as follows:

$$q_o(t) = K_o \omega(t),$$

where

$q_o(t)$ = the output flow rate in gallons/s
K_o = an output flow constant
$\omega(t)$ = the speed of the motor in radians/s

EXERCISES

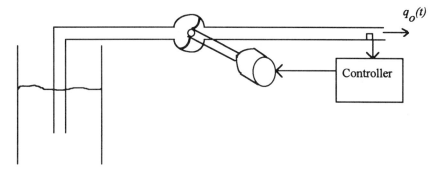

Fig. 5.31 Motor driving an impeller.

The torque present on the motor is almost entirely from the damping of the fluid.

(a) Develop the transform for the actuator plus process block.
(b) The following data pertains: $K_T = 10$, $K_B = 1$, $I = 2$, $R = 4$, $f = 0.1$, $K_o = 0.02$. Assume I am operating the system in open loop. I want an output flow rate of 0.5 gal/s. How many volts must I apply to the motor to obtain that flow rate in steady state?
(c) The transducer measuring the flow rate converts the rate of flow to a voltage by the following equation:

$$V_f = K_f q_o(t),$$

where:

K_f = the feedback constant in volts/gal/s
V_f = the feedback voltage

Using a control action K_c, solve for the closed loop transfer function of the system.

(d) Assume $K_f = 1.0$. Solve for the closed loop response to a step input of $q_d = 0.5$ and an integral controller with $K_i = 1.0$.

11. Figure 5.32 shows a fluid system in which the height of the fluid, H, is determined by the pipeline pressure, $P1$. The physics of the system is as follows:

(i) The flow rate into the tank is a function of the pressure drop across the inlet restriction. For the purposes of modeling this problem, it is assumed to follow the equation:

$$Q = \frac{P1 - P}{R}$$

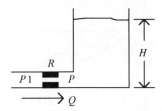

Fig. 5.32 Physical model for Exercise 11.

where:

Q = inlet flow rate
$P1, P$ = pressure
R = fluid resistance caused by the restriction (a constant).

(ii) The pressure at the bottom of the tank is:
$$P = \rho H$$
where:

ρ = density of the fluid (a constant)
H = height of water in the tank

(iii) The rate of fluid inflow determines the change in volume of fluid in the tank:
$$Q = A \frac{d}{dt} H$$
where:

A = cross-sectional area of tank

(a) Develop the open loop transfer function in the Laplace domain that describes the output/input relationship ($H(s)/P1(s)$) for this system.
(b) Assume the following data: $\rho = 1.0$; $A = 4$; $R = 0.5$. What is the open loop time domain response of a step input of pressure $P1 = 10$ and no water initially in the tank?
(c) What is the answer to part (b) if the initial height of water, $H(0) = 2$.
(d) Assume a controller is installed with unity feedback and a proportional control strategy where $K_p = 1$. Solve for the closed loop transfer function of the system assuming that the operator input is $H_d(s) = 4s^{-1}$ and $P1(s) = K_p e(s)$, and $H(0) = 0$.
(e) Assuming the system is to be run by a digital controller with a 1-second update period, develop the equations necessary to simulate the system and solve for the first three values of $H(s)$ in the simulation. Assume an initial condition, $H(0) = 0$.

12. For the thermal control problem of section 5.11.2, the following data applies: $R = 5$, $K_H = 4$, $K_p = 4$, $C = 2$, $T_a = 0$. What is the response of the system to a desired temperature, $T_d = 100$?
13. For Exercise 12, assume that the analog controller is replaced by a digital controller with a sample update time of 2 seconds. Using a simulation approach, as in section 5.12.1, simulate the behavior of the system over 10 update periods.
14. For the conditions of Exercise 13, develop an analytical model of the discrete controller and solve for the first four update periods.
15. Using the data of Exercise 12 and assuming an analog controller, what is the response of the system if the ambient temperature, $T_a = 50$ and the initial temperature, $T(0) = 50$?
16. For Example 5.17 of the text, develop the analytical model for a discrete controller. Using the data given in Example 5.17, proportional control, and a sample update time of 3 seconds, compute the system output for the first four update periods.
17. For Example 5.17 of the text, develop the analytical model for a discrete controller using integral control and an update time of 1 second. Using the data of Example 5.17, compute the system response for the first four update periods.
18. Table 5.2 of section 5.12.1 described a simulation in which the control action is updated immediately after sampling the process every Δt increments of time. Develop a table for $\Delta t = 1s$ and $\Delta t = 2s$ showing the levels of $h(t)$ and $\theta(t)$ for the case where Δt is the time interval between sampling the process and updating the control action.

FURTHER READING

Astrom, K. J. and Wittenmark, B. (1984) *Computer Controlled Systems: Theory and Design*, Prentice-Hall, Englewood Cliffs, New Jersey.

Bollinger, J. G. and Duffie, N. A. (1988) *Computer Control of Machines and Processes*, Addison-Wesley Publishing Company, Reading, Massachusetts.

Deshpande, P. B. and Raymond, H. A. (1988) *Computer Process Control*, Instrument Society of America, Research Triangle Park, North Carolina.

Electro-Craft Corporation (1980) *DC Motors, Speed Controls, Servo Systems*, 5th edn, Electro-Craft Corporation, Hopkins, Minnesota.

Groover, M. P. (1980) *Automation, Production Systems, and Computer Integrated Manufacturing*, Prentice-Hall, Inc., Englewood Cliffs, New Jersey.

Schwarzenbach, J. and Gill, K. F. (1984) *System Modelling and Control*, 2nd edn, Edward Arnold, London.

Thompson, S. (1989) *Control Systems Engineering and Design*, Longman Scientific & Technical, Essex, England.

6 Programmable logic controllers

6.1 INTRODUCTION

The programmable logic controller (PLC) is a special-purpose computer used to control the operation of electromechanical devices. It is special-purpose in the sense that it has been engineered for use in manufacturing environments, where electrical noise and mechanical vibration may be present, and it is programmed using a special-purpose language compatible with the requirements of sequential control of electromechanical systems. Any computer having input and output interfaces can be used to control external devices. However, several difficulties are immediately encountered when using an ordinary microcomputer for this purpose. The first is that most computers are not industrially hardened and are likely to be damaged when used on the factory floor. Currently, some vendors do offer industrially hardened general-purpose microcomputers and this difficulty is becoming less of a problem. A second difficulty is that the I/Os of general-purpose microcomputers were never engineered to handle line voltages and currents above TTL levels. In order to control an external device, e.g. a relay which energizes a circuit at 120 volts, the output port of the microcomputer must be electronically isolated to avoid destruction. PLCs have such isolation built into their inputs and outputs.

Finally, microcomputers typically offer a number of programming capabilities, including the use of several languages for writing code. When semiconductor logic functions became available and programmable factory controllers could be built, it was apparent that it would be necessary to standardize on a language. Without standardization it would be difficult for technicians to troubleshoot programs that had been developed by other technicians. The language that technicians had traditionally used to design and document factory control circuits was ladder diagramming of relay logic. In order to introduce the programmable controller into the factory with minimum retraining, the language of 'ladder logic' was adopted as the standard.

Prior to the late 1960s the factory control systems for a machine, a conveyor or other equipment was made up of a series of hard-wired contacts and relays that performed on/off operations in a specified sequence to control the desired work cycle of the machinery or equipment. These control panels took up considerable room and required a good deal of technician time to rewire when an application changed. By using a PLC, the relay logic formerly performed by the hard-wired contacts would be programmed as boolean equations in the memory of the PLC. This eliminated some of the cost of hardware and reduced a rewiring problem to a much simpler reprogramming problem.

6.2 PLC HARDWARE

Figure 6.1 shows the components of a typical PLC. As in the case of a microcomputer, there is a microprocessor, memory, I/O and a programming interface. Memory is usually battery-backed in order to avoid losing a program during power loss. The inputs and outputs are usually binary, i.e. single bit contacts that are read as an off/on (1, 0) condition in memory. These inputs and outputs are protected from damage by external line voltage levels by input/output circuits that isolate external voltages from the PLC processor. Figure 6.1 is typical of a small unitized PLC.

Some PLCs have the capability of being used in analog control and other advanced applications. Typically, this is done through the use of "analog modules" that are A/D and D/A converters designed to interface with the manufacturer's processor hardware. Each channel of a D/A or A/D converter will communicate with the processor through an assigned memory address of the processor. Figure 6.2 illustrates the general structure of a PLC processor with analog channels.

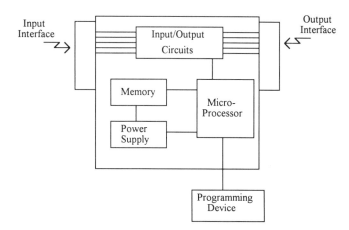

Fig. 6.1 Components of a typical PLC.

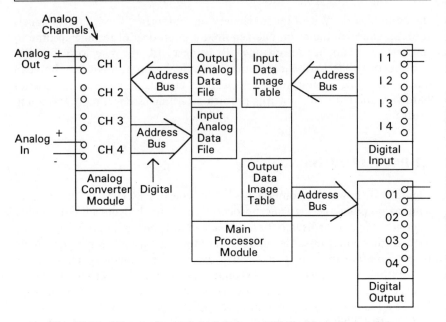

Fig. 6.2 Modular configuration of advanced PLC with analog module.

Finally, external communication via serial port is becoming an important feature of PLC design. PLCs were originally used in self-contained control applications, independent of other activities taking place in the factory. However, the trend today is to design factory control systems so that information and instructions can be communicated between PLCs and more powerful computers responsible for overall factory-floor control. For this reason, RS232 or local area network communications capabilities are available as part of PLC specifications.

Figure 6.3a shows a modular PLC. It is shown with a digital input module and a digital output module. The modules slide into a rack, the backplane of which provides the circuits connecting the modules to the processor. The backplane is an extension of the processor address bus and each module is an addressable portion of the PLC memory map. Figure 6.3b shows the PLC as it might be housed in a control cabinet on the factory floor. Operator communication and programming of the PLC is done using the PC interface above the PLC. Here the PLC is controlling the heating cycle of the retort that is partially visible on the left. Another typical configuration is to mount the PLC in a control cabinet with push buttons for inputing operator commands and digital readouts and lights for indicating status to the operator.

PLC HARDWARE

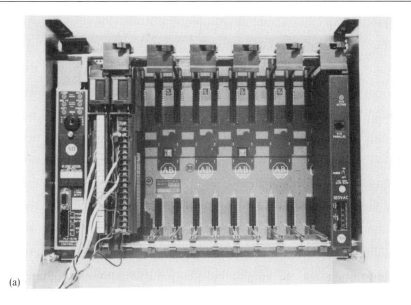

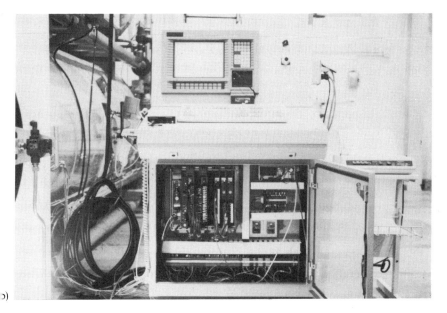

Fig. 6.3 Programmable logic controller and factory application. (a) Programmable logic controller; (b) PLC in control cabinet.

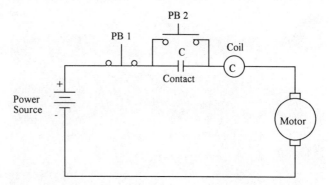

Fig. 6.4 Circuit diagram of motor on/off control.

6.3 RELAY CIRCUITS AND LADDER DIAGRAMS

As previously stated, the programming language for PLCs was a carryover from the diagrams used by factory technicians for wiring and troubleshooting hard-wired control cabinets. It is useful to examine a simple control application to understand the logic behind ladder diagraming.

Figure 6.4 is a circuit diagram for turning on and off a motor using momentary contact push button (PB) switches. PB1 and PB2 are normally closed and normally open push button switches, respectively. When PB1 is pressed, an open circuit is created at PB1; when PB1 is released, the circuit closes again. When PB2 is pressed, the circuit is closed at that point; when PB2 is released, the circuit reopens again.

Coil C and contact C operate together. Coil C is a relay coil, which consists of a core in an electromagnet. When the relay coil is energized, the core moves so that it closes the contact C. As long as the coil is energized, contact C will remain closed. When the coil is de-energized, the contact will reopen. A drawing illustrating a coil operating a contact is shown in Fig. 6.5. A single coil can operate one or more normally open and normally closed contacts.

The function of the circuit can now be evaluated. When the power source is initially turned on, the motor remains off because no closed path exists around the circuit. In order to start the motor, PB2 must be pressed. When PB2 is momentarily pressed, a closed circuit exists through PB1 and PB2, starting the motor and energizing coil C. Coil C is also referred to as a **control relay**. When coil C is energized, contact C closes. Coil C 'controls' contact C. When PB2 is released, the motor continues to operate because continuity around the circuit now exists through PB1 and contact C, which is now closed. The motor can be stopped by pressing contact PB1 momentarily. This action opens the circuit, de-energizing coil C and, consequently, opens contact C. The motor operation ceases until restarted using PB2. This simple illustration demonstrates some of the symbols used in factory wiring diagrams and the logic of their operation.

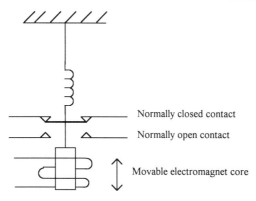

Fig. 6.5 A coil operating a set of contacts.

Figure 6.4 is a diagram of a linear circuit. However, factory control diagrams follow a different format called **ladder logic** diagramming. The correspondence between a linear circuit and a ladder diagram can be seen in Fig. 6.6. The uprights of the ladder correspond to power and ground of the power supply, i.e. they represent a voltage differential that provides power to the circuit when continuity is established across the rung. The rungs of the ladder are used to represent the circuit. A convention employed for ease of understanding is to use one rung of the ladder for each logic function required to describe the functioning of the circuit. The boolean logic for the circuit of Fig. 6.6 is shown to the right of each rung. Coil C is energized if $PB1 \cdot (PB2 + C)$ is true. Energizing coil C turns on the motor. It would be incorrect to program the logic on one rung as $MOTOR = PB1 \cdot (PB2 + C)$. This would be incorrect because C must be retained in the enabled state before the motor will run. While PB1 and PB2 are enabled by an operator, the above equation does not describe the condition for setting C. If C is initially open and the motor is off, $MOTOR = PB1 \cdot (PB2 + C)$ states that pushing PB2 will momentarily start

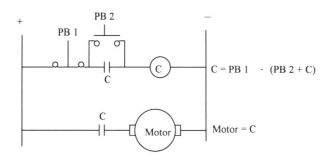

Fig. 6.6 Ladder diagram of Fig. 6.4.

the motor, but the motor will stop when PB2 is released. Hence, the logic is incomplete. Using the two ladder rungs of Fig. 6.6, which imply the corresponding boolean equations, the logic for operating the motor is complete.

This brings us to the function of the programmable logic controller. If Fig. 6.6 represents a hard-wired circuit, we can see that there are three hardware components necessary to operate the motor: two push button contacts (PB1 and PB2) and one control relay (C). Of the three components, only two need to be accessible to an operator, specifically PB1 and PB2. The other component is internally operated according to the logic given to the right of Fig. 6.6. Since logic operations can be programmed on a computer, we can eliminate the coil and contact by programming their operation as boolean equations on a computer.

Figure 6.7 is a block diagram showing the equivalent operation on a PLC. The two push button contacts are wired as inputs to the PLC and the motor is wired as an output. The program consists of two boolean equations, which come from the ladder diagram of Fig. 6.6. The PLC operates by continuously looping through the program, evaluating each equation sequentially. If the conditions on the right-hand side of the boolean equations are true, the memory location of the binary variable on the left hand side is set (1) accordingly. If the argument is not true, the binary variable is reset (0). The execution sequence for turning the motor on and off is shown in Table 6.1. This corresponds to the logic given in Fig. 6.7.

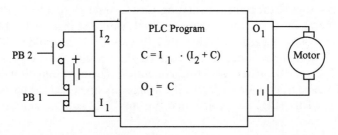

Fig. 6.7 PLC input/output wiring and program to perform function of relay circuit of Fig. 6.4.

Table 6.1 State Table for Fig. 6.7

PB 1	PB 2	C	Motor
1	0	0	0
1	1	1	1
1	0	1	1
0	0	0	0

6.4 SEQUENTIAL CONTROL USING RELAY CIRCUITS

When sequential control systems were hard-wired, a standard set of symbols were developed to document the control logic. The most commonly used of these symbols are shown in Fig. 6.8. For the purpose of description, these symbols are grouped as follows: coils and their contacts, external contact devices, and activated devices.

The control relay has been described already. When the coil is energized, the relay causes its associated contact to change from its normal state to its complement state. Therefore, when energized, a normally open contact will be closed and a normally closed contact will be open. When the control relay is de-energized, its associated contacts revert back to their normal state.

Contacts that are coil-operated are typically part of the internal electrical control logic. There are other contacts that are mechanical closures operated by forces outside the internal electrical control logic. The push button

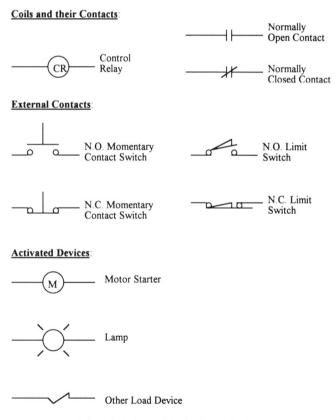

Fig. 6.8 Relay ladder logic symbols.

momentary contact switch is used primarily for initiating or terminating system operation. An example is a start button on a machine that a machine operator would use to initiate the machine operating cycle. The limit switch, described in Chapter 3, is a sensor that is usually used to let the control circuit know that the mechanical device it is operating has reached a particular point in its operating cycle. When that point is reached, the force of the mechanical device closes or opens the limit switch. If the switch is normally open, the force will close it; if it is normally closed, the force will open it. When the force is removed, the limit switch will revert back to its normal state.

The control circuit is used to drive actuators, or activated devices. The motor starter symbol indicates a motor as an activated device. It is also typical to have lamps (lights) being activated during system operation. The light would be mounted on a control panel and indicate systems status, e.g. an 'ON' light to indicate the machine was powered up. Other load devices are indicated using the symbol as shown.

The use of these symbols in documenting a control circuit is illustrated in the following example.

EXAMPLE 6.1

Figure 6.9 shows a drilling machine for which we would like to design a control circuit. The machine is operated as follows. Power is brought to the machine by pressing the 'power on' button. This provides power to operate the actuators and power the rails of the ladder, thus activating the relay ladder logic. The operator places a workpiece on the fixture of the drill press. The work cycle begins when the operator presses the 'start cycle' button on the control panel. When the operator does this, the drill head begins to descend and a pneumatic clamp is activated to hold the work piece in place. Simultaneously, the drill spindle motor is activated and drills a single hole in the workpiece. Drilling is complete when the drill head reaches a limit switch (LS2) mounted on the Z-axis column. At that point the drill head is retracted up the Z-axis column until it reaches a top limit switch (LS1). The operation is then complete, the fixture clamp is released, and the operator removes the workpiece. The work cycle will begin again when the operator loads a new workpiece and presses the 'start cycle' button.

The spindle motor that turns the drill and the Z-axis motor that lowers and raises the drill head are both operated by closing and opening contacts in the motor drive circuit. The pneumatic cylinder that clamps the workpiece is a single acting pneumatic cylinder driven by a solenoid, as shown in Fig. 6.9. When electrical power is supplied to solenoid A, the cylinder is driven to its fully extended position. For this position the workpiece is clamped. The cylinder will change position only when power is removed from solenoid A and the spring returns the cylinder to its rest position, i.e. the workpiece will become unclamped. Design a relay circuit to perform this work cycle.

SEQUENTIAL CONTROL USING RELAY CIRCUITS

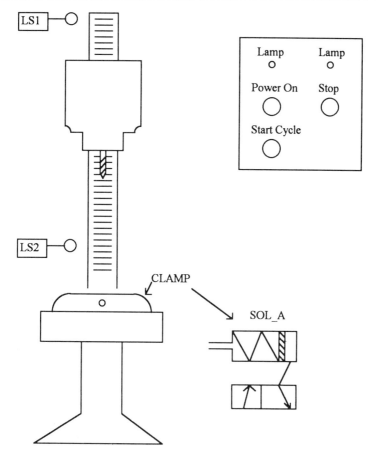

Fig. 6.9 Drill press operation.

Answer:

Before we proceed with the design, the input and output devices should be listed. They are as follows:

- Inputs:
  PB3 Start cycle momentary contact push button (normally open).
  LS1 Upper travel limit switch (normally open).
  LS2 Lower travel limit switch (normally open).
- Outputs:
  SOL_A Solenoid valve that operates the clamping device.
  M1   Spindle motor.
  M2   Negative Z (downward) travel for drill head motor.
  M3   Positive Z (upward) travel for drill head motor.

RUNG

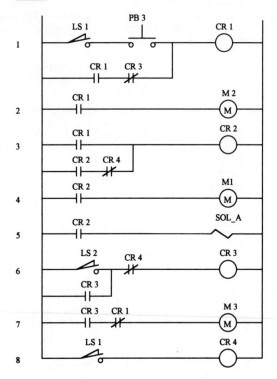

Fig. 6.10 Work cycle control circuit.

Figure 6.10 shows a relay ladder diagram that represents one solution to this problem. The following is a description of the operation of each rung of the ladder.

- Rung 1: At the start of the cycle, the drill head should be at its uppermost position, closing limit switch LS1. Hence, pressing PB3 will activate control relay CR1. This rung will be retained in the enabled state as long as CR3 does not open.
- Rungs 2, 3: Enabling CR1 starts the downward Z drive (M2) and enables CR2.
- Rungs 4, 5: CR2 controls the spindle motor (M1) and the solenoid clamp (SOL_A). These outputs are being controlled separately from the $-Z$ drive because we wish to keep them on when the $-Z$ drive is turned off and the drill head is being raised.
- Rung 6: When the drill head reaches full travel in the negative Z direction, LS2 will close, enabling CR3. Note that this action opens contact CR3 in rung 1, disabling CR1. This, in turn, disables the $-Z$ motor control (M2) in rung 2.

SEQUENTIAL CONTROL USING PLCs

- Rung 7: Enabling CR3 turns on the +Z motor control if CR1 is de-energized. It is safe practice to declare in the logic of the rung that you will not permit the +Z and −Z drives to be enabled simultaneously.
- Rung 8: At full +Z travel, LS1 will be closed, enabling CR4. This turns off M3 (rung 7) by opening contact CR4 (rung 6). This also stops the spindle motor (rung 4) and unclamps the workpiece (rung 5) by opening CR4 (rung 3).

At this point the cycle is complete. It can be started again by pressing PB3.

6.5 SEQUENTIAL CONTROL USING PLCs

As described in section 6.3, a programmable logic controller can be used to replace the internal control logic of a relay control circuit. The example of Fig. 6.10 has three input devices and four output devices; the remainder of the circuit is composed of switching devices that perform the control logic. These devices could be boolean statements in the memory of a computer that performs the switching logic. Furthermore, because the computer is reprogrammable, such an implementation will allow the user to alter the switching logic if the application changes. In this section we discuss the use of a PLC in lieu of relay control circuits.

The logic concepts used in relay ladder logic have been adopted for the programming of PLCs. Symbols have been simplified and they differ somewhat among PLC manufacturers; however, the differences are not very great. Rather than describe several symbolic representations, we shall use a representative set of symbols and illustrate their use. These symbols are shown in Fig. 6.11.

In general, rounded brackets are used to represent entities that are energized, whether they are internal control entities, such as binary bits that replace coils, or externally controlled outputs, such as motors, solenoids or lamps. The particular device is identified by either a number or a letter/number combination. In Fig. 6.11, the control relay bit is indicated by the letter B followed by a numerical designation. The control relay bit operates one or more contact bits. The normally open contact bit B10 is controlled by control relay bit B10.

The bit number is assigned by the programmer of the ladder diagram. For each bit number available to be used by the programmer, the manufacturer of the PLC has designated an address location in the processor memory that will retain the current status of the bit. For example, if the processor is an 8-bit microprocessor with associated memory, the manufacturer will assign an 8-bit address to hold the status of bits B0 through B7. The next 8-bit address will be assigned to retain the status of B8 through B15, and so forth. The programmer is not concerned with how the bits are assigned to processor memory. The programmer simply uses the bits as a substitute for physical coils and contacts

Coils and their Contacts:

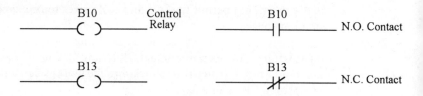

External Inputs:

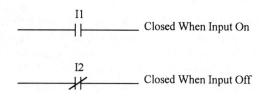

Activated Output Devices:

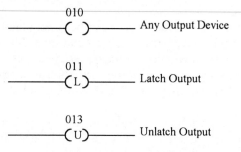

Fig. 6.11 Basic PLC ladder logic symbol.

and arbitrarily selects which bits to use from the range of bits made available in the PLC.

In Fig. 6.8 different symbols were used to indicate different types of external contacts that served as inputs to the control circuit. This practice was used in relay diagrams in order to properly identify the kind of device that is wired into the control cabinet. From the point of view of PLC software, it does not matter whether the external device it is monitoring is a push button switch, limit switch or some sort of sensor. Once the control circuit is reduced to boolean algebra, the programmer is only concerned with knowing that the program must monitor the state of an input signal at a specific rung in the ladder. For that reason, external input devices are modeled simply as normally open and normally closed contacts. They are identified as inputs by the prefix I. In

SEQUENTIAL CONTROL USING PLCs

Fig. 6.11 we have indicated I1 and I2 as input signals 1 and 2, respectively. Input signal 1 is enabled when the contact is closed, i.e. the PLC will consider I1 true when it reads a positive voltage on input channel 1. On the other hand, I2 is closed when the PLC reads zero volts on input channel 2 and open when an input voltage is present.

As in the case of input devices, most PLC programming languages do not distinguish between specific actuated (output) devices. From the point of view of the PLC, any digitally activated device is just another output that can be turned on or turned off by setting or resetting the output control register that controls it. As shown in Fig. 6.11, output devices are indicated by the letter O preceding the output channel designation.

A special case of setting an output is the LATCH and UNLATCH, indicated by L and U within the rounded brackets. When an output is 'latched', it will remain enabled even if the contact bit closing its rung is reset. Once set, the output will remain set until it is unlatched. This is done by enabling a rung having an 'unlatch' coil with the output designation of the previously 'latched' output that is now being 'unlatched'. Therefore, the latch output and unlatch output must work together in controlling the ON/OFF state of a specific latched output.

Figure 6.11 illustrates the most common PLC programming symbols. The following example indicates how they can be used.

EXAMPLE 6.2

Using the relay ladder diagram of Fig. 6.10, program the control software for a PLC.

Answer

We first note the number of inputs and outputs involved in this problem and assign them to input and output addresses of the PLC. They are as follows:

	Relay circuit designation	PLC input/output designation
Push button 3	PB3	I3
Limit switch 1	LS1	I4
Limit switch 2	LS2	I5
Spindle drive	M1	O1
$-Z$ motor drive	M2	O2
$+Z$ motor drive	M3	O3
Clamp solenoid	SOL_A	O4

Using the PLC requires three input addresses and four output addresses. The coils and contacts can be eliminated since they are replaced with internal addresses (bits) in the PLC. A block diagram of the PLC and its connections is shown in Fig. 6.12. In Fig. 6.13 we have taken the relay ladder of Fig. 6.10 and developed the corresponding PLC program. It is left to the reader to compare the two ladder diagrams.

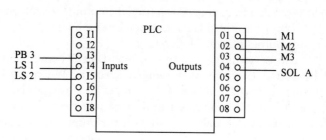

Fig. 6.12 Block diagram of I/O wiring to PLC.

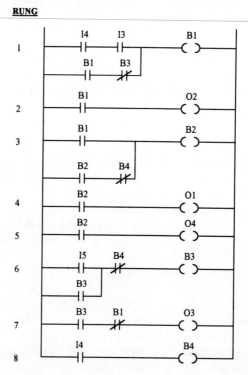

Fig. 6.13 PLC program corresponding to relay circuit of Fig. 6.10.

6.5.1 Further simplification of ladder logic

Figure 6.13 exactly mirrors Fig. 6.10; the only change is the symbology. However, underlying Fig. 6.10 is the requirement to have a coil associated with a contact in the internal logic of the control system. Thus, in Fig. 6.10, in order to operate M2, CR1 coil must close CR1 contact and retain itself in the enabled state. However, in PLC software, this function is performed by data bits. An output of a PLC, e.g. O1, is both a physical output and a data bit. The stored user program sets data bit O1 when the conditions on the rung are true. When data bit O1 is set, the output circuits of the PLC enable power to the physical device by switching the appropriate voltage of the output channel. Since O1 is a data bit, it can be used for internal control in the same manner as any other bit in the memory of the PLC. In effect, a more parsimonious version of Fig. 6.13 can be obtained by using output bits as information bits for internal retention of ladder rungs.

The first PLCs required the use of logic exactly analogous to relay circuits because technicians had been trained to think of the PLC as a direct replacement for the relay circuit. As users became more computer literate, the programming discipline began to reflect the fact that the PLC was just a special-purpose computer. In the next example, we illustrate this transformation.

EXAMPLE 6.3

Convert Fig. 6.13, using output bits to retain rungs where possible.

Answer

Figure 6.14 shows the reduced ladder logic diagram. Two rungs have been eliminated. The logic for the change can be seen by comparing rungs 1 and 2 of Fig. 6.13 to rung 1 of Fig. 6.14. In Fig. 6.14 the O1 bit is used both for output and to retain itself. Hence, in many cases it is possible to eliminate the indirect retaining of outputs that was necessary in relay circuit implementations. Henceforth in this chapter we shall use the conventions of Fig. 6.14.

6.5.2 Program design and fail safe condition

Aside from the question of programming the logic necessary to operate a machine, the engineer must also consider the safe operation of the program should an unanticipated failure occur in the system. At the time a program is being created, it is not always obvious how a combination of events can lead to an unsafe condition. However, there is a standard precaution that is always taken. It can generally be described by the rule that a recognized potential

RUNG

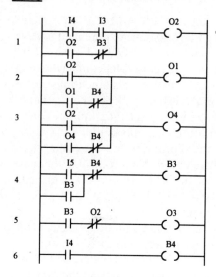

Fig. 6.14 Reduction of Fig. 6.13.

failure of the system should be addressed such that, should the failure occur, the system will fail in a safe condition. The term 'safe' refers to not endangering the operator and equipment.

An example of a potential failure occurs in the program of Fig. 6.14. In that example, normally open contacts were used for the upper and lower limit switches. We are depending on the closure of these contacts to signal the fact that the machine has reached the limit of its travel and the motor drive for that direction should be turned off. Consider the possibility that a break or loose connection occurs in the wiring from the limit switch to the controller. Should that occur, the controller will always read the limit switch as being in the disabled condition. When the limit switch is engaged by the drill head, the motor drive will not be turned off. The continued operation of the motor beyond the appropriate limits of travel could cause damage to the machine and, possibly, the operator.

One solution, of course, is to add redundancy to the system by installing two upper and two lower limit switches. The program should allow the closing of either limit switch to disable the motor drive. This will prevent an accident if one of the limit switches of each pair fail.

Another solution is to use normally closed contacts for the limit switches. A normally closed contact, when it is not enabled, will mean that the drill head is within the travel limits. The controller will read a positive voltage level when the drill head is within the travel limits because the contacts are closed. When the travel limit is exceeded, the limit switch will open and the controller will

SEQUENTIAL CONTROL USING PLCs

read 0 volts. At that point the drill head has opened the contact and the motor drive should be disabled. With this arrangement, should a loose connection or wiring failure occur, the controller will assume that the contact has been opened. The motor will be disabled by the controller. No damage will be done. An operator, seeing that the machine does not respond to the start cycle button, will check for a loose wire as part of the troubleshooting procedure.

The logic for a program with this design will be different to that of Fig. 6.14. Designing such a program is left as an exercise for the reader at the end of this chapter.

6.5.3 Ladder logic and boolean algebra

By now the reader should recognize that ladder logic is just another formalism of boolean functions. In this section we will briefly touch on this point by looking at the structure of a ladder representation of common boolean functions. These are shown in Fig. 6.15.

- Equivalence: Fig. 6.15a illustrates equivalence, which is given by the boolean $A = B$.
- Complement: Consider two boolean variables that are complements, $A = \bar{B}$. Figure 6.15b is the ladder logic implementation.

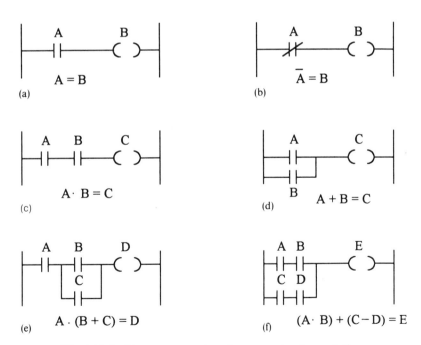

Fig. 6.15 Ladder representation of common boolean relations.

- Logical AND: The logical AND is the simple serial rung, as shown in Fig. 6.15c, where $A \cdot B = C$.
- Logical OR: The logical OR is the parallel implementation of rungs, as shown in Fig. 6.15d, where $A + B = C$.
- Combined logical operations: In many cases ladder rungs will require combined logical operations. In Fig. 6.15e, $A \cdot (B + C) = D$. In Fig. 6.15f, $(A \cdot B) + (C \cdot D) = E$.

EXAMPLE 6.4

Write the boolean expressions for the ladder of Fig. 6.14.

Answer

Rung 1: $(I4 \cdot I3) + (O2 \cdot \overline{B3}) = O2$
Rung 2: $O2 + (O1 \cdot \overline{B4}) = O1$
Rung 3: $O2 + (O4 \cdot \overline{B4}) = O4$
Rung 4: $(I5 + B3) \cdot \overline{B4} = B3$
Rung 5: $B3 \cdot \overline{O2} = O3$
Rung 6: $I4 = B4$

6.6 SCAN SEQUENCE AND PROGRAM EXECUTION

Section 6.5 should give the reader a hint about how PLC programs are executed. In this section we shall discuss the execution cycle in a little more detail.

Figure 6.16 illustrates the sequence of events executed by the processor in running a PLC program. The term 'scan' is used to refer to one complete execution cycle. It is composed of two subcycles: I/O scan and program (ladder) scan.

When an output of a program is enabled, a memory location associated with that output bit is set (becomes true). This output memory location resides in the 'output image table'. Before the actual output port can be enabled or disabled, the state of the output image table must be copied to the output port. Similarly, when an input is turned on, it registers immediately at the input port. However, before it is used for program control, the status of the input port is copied into memory. This memory is referred to as the 'input image table'.

When the scanning cycle begins, the I/O scan is performed first. During that scan the status of the output image table is copied to the output port and the status of the input port is copied into the input image table. Thus, the I/O image table represents current information at the time of the I/O scan.

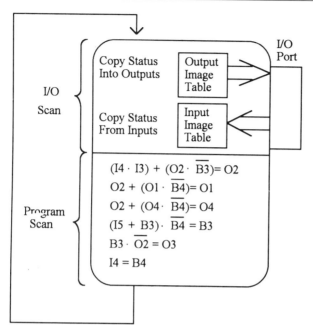

Fig. 6.16 PLC program execution sequence.

The program scan reads sequentially each rung of the ladder. It evaluates each equation based on the current status of the I/O image table and sets the results of the expression accordingly. When the program scan is complete, the I/O scan begins again with the new output image table. A PLC executes this cycle over and over again, updating inputs and outputs, and re-evaluating ladder rungs.

6.7 DECOUPLING AND PROGRAM CONTROL

As explained in section 6.6, the PLC executes the program by scanning the entire program during each scan cycle. This differs from programming in high-level languages, where program control is exercised by having a line of code direct the program to the next line of code or subroutine to be executed. This happens with the use of unconditional and conditional branching instructions, such as 'if...then...else'.

Despite the fact that each line of code is read during each scan, a good PLC program should be structured to isolate sections of a program, structuring it in a manner analogous to that of a structured programming language, such as

Pascal or C. In fact, it is often necessary to do so when two or more sections of the program must be decoupled from each other. In this section we discuss the need for structuring a ladder logic program by illustrating an example in which decoupling of program subroutines is required.

EXAMPLE 6.5

We refer again to the drilling machine in Fig. 6.9. Before an operator can use the machine in its normal work cycle, it must go through its start-up cycle. This begins when the operator presses the 'power on' button on the panel. When this is done, power is brought to the machine and to the ladder logic program. When power comes on, we want the program to turn on the 'power on' lamp and to take the drill head upward until LS1 is engaged. This is the 'home' position of the machine, the position from which it can begin its work cycle. When the home position is reached, the 'ready' lamp should light, indicating to the operator that he or she may proceed with the work cycle.

At any time during the work cycle or homing cycle, the operator can intervene in the operation by pressing the 'STOP' button, which stops the operation of the machine by removing power to the machine. If this is done, the operator must reinitialize the machine by going through the start-up and homing procedures again.

Prepare the ladder logic diagram in order to control all operations of the machine.

Answer

For this situation there are actually two subroutines required; one for the homing cycle at power up and one for the work cycle. When the homing cycle is complete, it is desirable to isolate the homing subroutine, decoupling it from the work cycle so that it cannot become active again until after the stop button is pressed. The new controls to be considered are as follows:

Input	PLC Input Designation
Power on button	None
Stop button	None
Start cycle button	I3
Limit switch #1	I4
Limit switch #2	I5

DECOUPLING AND PROGRAM CONTROL

261

Output	PLC Output Designation
Spindle drive	O1
$-Z$ motor drive	O2
$+Z$ motor drive	O3
Clamp solenoid	O4
Power on lamp	O5
Ready lamp	O6

Note that neither the power on button nor the power off (stop) button are part of the PLC program input. These are hard-wired contacts that bring power to the machine via a relay circuit. Such a circuit is implemented using control relays as discussed in section 6.4. It is also safe practice in machine design to use 'hard stops' (called E-stops) for shutting a machine down quickly during its cycle when an unsafe condition is observed. These requirements can be implemented so that they turn off power to the actuators, but not to the processor. However, the processor must be informed of the 'off' condition of an actuator so that it updates its output image table and maintains correct knowledge of the state of the system. This can be done by wiring the **master control relay** (MCR), which is opened and closed by the power on and stop buttons, as an input to the PLC. The MCR input can be programmed as an input contact in each rung of the control program that controls an output. When power to the system is turned off, each output of the program is disabled by opening the MRC. This practice will be assumed in this chapter, but will not be made explicit in our examples, which are being used to illustrate control of operating cycles only. The reader is asked to explicitly implement this additional feature in end-of-chapter exercises 4 and 6.

'Soft stops' (software stops) may also be employed if it is deemed appropriate, but this is in addition to a hard stop, sometimes referred to as an E-stop, or emergency stop. The E-stop should remove all power from the machine immediately, eliminating any further motion from taking place.

Figure 6.17 illustrates a program that is capable of performing the required functions. This figure also illustrates another important point, namely, good documentation practices. We have annotated each contact and output with the title of its function. If the reader compares Fig. 6.17 to Fig. 6.14, it will become obvious that the interpretation of the ladder has been greatly improved by labeling the elements.

The first three rungs control the start-up cycle. The interpretation is as follows:

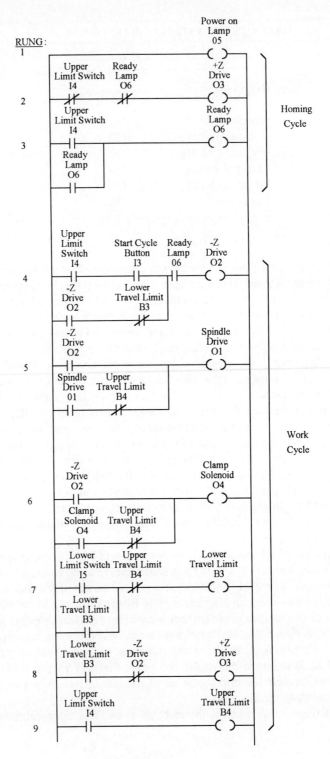

Fig. 6.17 Program of Example 6.5.

COUNTERS AND TIMERS

- Rung 1: Pressing the power on button, which brings power to the system, enables the power on lamp, which will stay on as long as there is power to rung 1 of the PLC.
- Rung 2: If the upper limit switch is not enabled and the ready lamp is not on, the +Z drive will be turned on and the drill head will move to its home position.
- Rung 3: When the drill head enables the upper limit switch, the ready lamp (O6) will go on. This disables the +Z drive in rung 2 and rung 2 will not execute again unless the ready lamp is disabled.
- Rung 4: The addition of the ready lamp in this rung prevents the start cycle button from initializing a work cycle until the homing cycle is complete.

Note that it is possible that, upon start up, the drill head may already be in the home position. In that event, I4 is enabled and rung 2 will not execute. The ready lamp will go on immediately.

Rungs 4–9 are analogous to rungs 1–6 of Fig. 6.14. The only difference is the insertion of contact O6 in rung 4 of the work cycle circuit to ensure that the work cycle cannot proceed until the start-up cycle is completed. In effect, we are using O6 to decouple rungs 1–3 from rungs 4–9. Furthermore, if the stop button is pressed, this will remove power from the system by opening MCR and disabling all outputs, including O5 and O6. The only way to bring the machine back up is to enter the start-up cycle again. In this manner a structure has been added to the program which decouples the start-up cycle from the work cycle. It is left to the reader to go through rungs 4–9.

6.8 COUNTERS AND TIMERS

Another set of functions available in PLCs are counters and timers. The manner in which they have been implemented and the symbology used differs somewhat among manufacturers. We shall describe a symbology similar to that used by the Allen-Bradley Corporation, which is the largest American manufacturer of PLCs.

6.8.1 Counters

A counter is a register in the PLC that accumulates a record of the number of times an event occurs. The counter may be used to keep track of events or it may be used to initiate an action once the counter has reached a certain value. Figure 6.18a illustrates a count-up counter in a ladder program. The counter is identified as counter C2. The 'Preset' register holds the value of the count at which the rung becomes true. The 'Accum' register holds the value of the actual accumulated count. The accumulator indexes by one with each transition of I1 from false to true. For example, I1 could be a proximity sensor sensing the passage of parts along a conveyor belt.

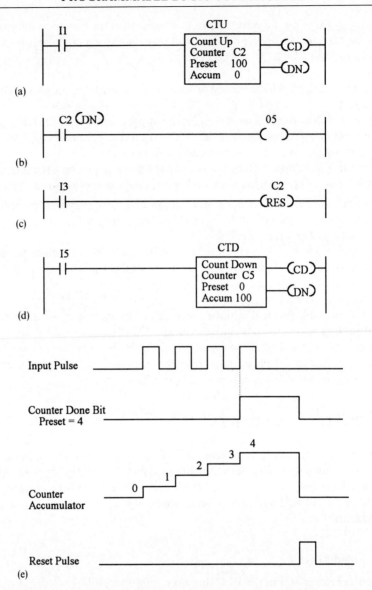

Fig. 6.18 Count-up and count-down counters and timing diagram for count-up.

When 'Accum' = 'Preset', the rung becomes true, i.e. the counter is done (DN). This sets a bit C2(DN) in the memory of the processor. Figure 6.18b shows another rung of a program in which counter C2(DN) bit is enabling an output when the counter has reached its preset value. The combination of Fig. 6.18a and 6.18b allows the counter to control an action in the program.

COUNTERS AND TIMERS

The 'Accum' value in a counter can be reset at any time by using a reset rung. Figure 6.18c shows a reset rung where counter C2 will have its 'Accum' go to zero whenever I3 becomes true.

An alternative to the count-up counter is the count-down counter shown in Fig. 6.18d. The count-down counter subtracts one from its accumulator for each transition of the rung condition. Both the Accum value and the Preset value are programmed. If a count-down counter is reset, the accumulator register reverts to its original value.

Figure 6.18e shows the timing diagram for a count-up counter. The counter accumulator increments at the leading edge of each input pulse; when Accum = Preset, the DN bit is set. The reset pulse clears the accumulator and disables the DN bit.

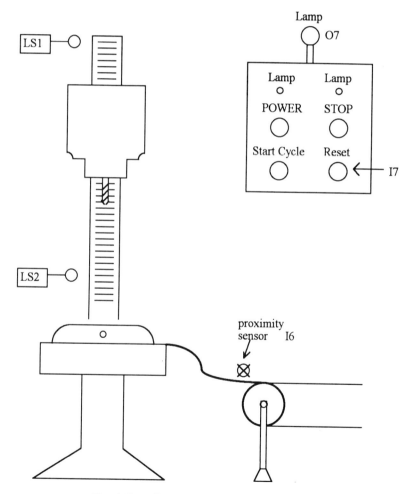

Fig. 6.19 Drilling operation with unit counts.

EXAMPLE 6.6

Figure 6.19 shows a continuation of Fig. 6.9. A discharge conveyor has been added to the side of the drill press. When the machine completes a work cycle and the workpiece is unclamped, the operator simply pushes the workpiece off the machine bed and onto the conveyor, from whence it is conveyed to the next operation.

The factory produces several parts that require drilling before going on to the next operation. Parts are manufactured in lot sizes of 100. When the machine operator has drilled 100 of a particular part, the fixture is to be reset so that production of 100 units of another part can begin.

It has been decided that the PLC program should keep track of the number of parts produced and alert the operator that a batch of size 100 has been reached. How can this additional requirement be handled?

Answer

There are many ways of implementing such a requirement. Counting the number of times the machine is put through a work cycle might seem like a reasonable way of keeping count. However, some work cycles may be executed without a part in the fixture. It is, in fact, the completion of a part that is the important event to be counted.

One implementation is shown in Fig. 6.19. Here we have placed a proximity sensor on the discharge conveyor to sense the passing of a part. This is wired as input 6 (I6) to the PLC. In addition, a red lamp has been placed atop the machine control panel as an output (O7) of the PLC. Finally, a push button input (I7) has been added as a reset button for the counter.

These new I/Os will be used as follows. When the count of parts reaches 100 and the drill head reaches its home position, the red lamp will be lit and the machine will be stopped. The operator adjusts the fixture for the new part, clears the accumulator of the counter by pressing the reset button, and begins the work cycle of the machine by pressing the start cycle button.

Figure 6.20 shows the new rungs added to the ladder of Fig. 6.17. It also shows how existing rungs are affected. An explanation is as follows:

- Rung 3: B1 has been added, which is enabled when 100 units are complete and the drill head is in the home position, thus preventing the operator from starting a new work cycle. B1 is controlled from rungs 10 and 11.
- Rung 10: The passing parts toggle the proximity sensor, increasing Accum by one each time. When Accum = 100, C1(DN) is set.
- Rung 11: C1(DN) enables B1 when the drill head has reached its home position. This disables the work cycle by disabling O6 in rung 3.
- Rung 12: Enabling B1 turns on the red lamp.
- Rung 13: Pressing the reset button sets Accum = 0, which disables C1(DN) (rung 10), which disables B1 (rung 11), which turns off the red lamp (rung 12),

RUNG:

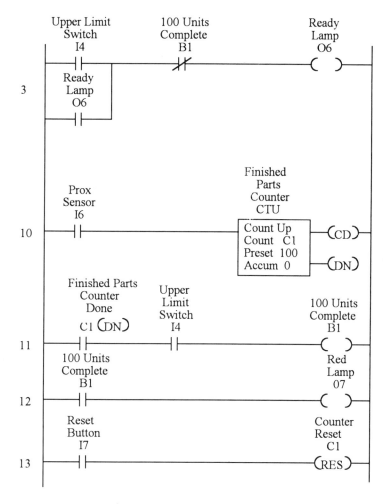

Fig. 6.20 Program of Example 6.6.

and puts the machine program in position to have its work cycle initiated (rung 3).

We note that the above program did not disable power to the machine. Instead, we used a software stop to prevent a work cycle from being initiated while the machine operator was adjusting the fixture. This was done for the purposes of illustrating a software stop. A much safer practice that is always followed in the design of machine control software is to remove power entirely from the machine when an operator is adjusting tooling or fixtures. This ensures that operator injury will not occur from an unforeseen program error. The reader will be asked to evaluate this in an exercise at the end of this chapter.

6.8.2 Timers

A timer accumulates the length of time that the precondition of its rung is true. Figure 6.21a illustrates a timer-on (TON) delay timer in a ladder rung. The

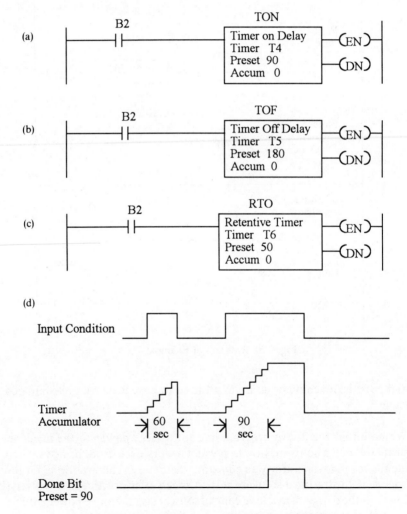

Fig. 6.21 Timers and timing diagram for TON timer.

COUNTERS AND TIMERS

TON timer begins to accumulate time when its input condition, in this case B2, becomes true. As long as the rung condition is true, the timer will continue to increment until it reaches its preset value. When 'Accum' = 'Preset', the done bit (DN) is enabled. As in the case of the counter, this can be used to control another rung of the ladder. Should the input condition, B2, become false before the timer reaches its Preset, the accumulated time will go to zero.

There is also a timer-off (TOF) delay timer. This timer is enabled when the rung condition goes from true to false. As shown in Fig. 6.21b, the timer accumulates time as long as B2 is false. When 'Accum' = 'Preset', the DN bit is enabled. Should B2 become true before the Preset value is reached, the accumulator will be reset.

The TON and TOF timers are non-retentive, i.e. if their rungs go false, the accumulated value is reset. In some cases it may be desirable to accumulate the total time for which an input condition was true, even if that time is not continuous. For this we require a retentive timer, i.e. a timer whose accumulator is not reset each time the rung condition is disabled. Figure 6.21c shows the format of a retentive timer. The format of a retentive timer is similar to a timer-on delay.

As in the case of counters, a timer can be reset at any time using the reset rung.

EXAMPLE 6.7

We return to Example 6.6 and Fig. 6.19. The discharge conveyor has a limited capacity to hold workpieces between operations. Occasionally workpieces may back up and fill the conveyor. It is also possible that workpieces may get hung up along the conveyor, causing a back-up to occur. It is desirable to alert the machine operator to these conditions so that he or she can investigate the problem.

Plant management is also concerned with minimizing the amount of time that the machine is not working on drilling parts. Of particular concern is the time it takes to change over the tooling and fixtures between groups of 100 parts. Management would like to collect data on the ratio of changeover time to run time on all the machines in the shop, including the drill press. What additions could be made to the system and software to accomplish these new requirements?

Answer

There are numerous ways of implementing these new requirements. Here we suggest some simple changes that minimize the amount of additional hardware.

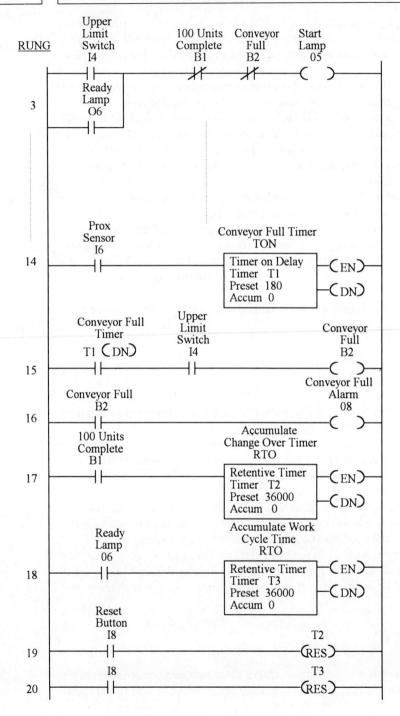

Fig. 6.22 Solution to Example 6.7.

COUNTERS AND TIMERS

The first requirement is to detect a back-up on the conveyor. Since we already have a proximity sensor on the discharge conveyor (I6), we can use it for this task as well. If the workpieces back up, the proximity sensor will be blocked by a workpiece on the discharge conveyor. Therefore, we can use this to enable a TON timer with a preset value of a few minutes. If the back-up clears, TON will be reset. If it does not clear within a few minutes, we stop the system and output an alarm, which will be added as O8.

The second requirement is to collect data on total changeover time and total work cycle time. In Example 6.5, we have established indicators of when the machine is in the changeover mode. We will use these indicators in conjunction with retentive timers to collect data on a daily basis. At the end of each day the accumulated value in these timers can be read out of the PLC and the timers can be reset for the next day's operation.

The ladder describing the additions required is shown in Fig. 6.22. The following is an explanation of each rung.

- Rung 3: Here we have added B2, which will shut the work cycle down if the conveyor is full or backed up.
- Rungs 14, 15 and 16 accommodate the first requirement, i.e. to indicate a full or backed up conveyor.
- Rung 14: This rung contains the timer-on delay, which will begin timing when the prox sensor has a part in front of it. If the timer is on for 180 seconds, the DN bit will be set.
- Rung 15: The DN bit sets B2, disabling rung 3 when the drill head has reached its uppermost position.
- Rung 16: When B2 is enabled, the conveyor full alarm will sound.

Note that the conveyor full timer, T1, will not disable the work cycle until the drill head is in the home position. It is undesirable to stop the machine in an unknown position. For example, it would be troublesome if the machine stopped while the drill was engaged in drilling the metal. Rungs 17, 18, 19 and 20 accommodate the second requirement, i.e. the collection of data on changeover time and work cycle time.

- Rung 17: When B1 is enabled, the changeover cycle begins. When this cycle ends, the operator presses push button I7 (see Fig. 6.20), which has the effect of resetting the counter and disabling B1. Since T2 is a retentive timer, it will collect the total accumulated changeover time for the day. A reset value of 36 000 seconds (10 hours) ensures that the timer can accumulate over an entire shift without timing out. The purpose of this timer is not to time out, but only to accumulate data.
- Rung 18: The machine is in its work cycle whenever O6 is enabled. Rung 18 is analogous to rung 17, except it is collecting total work cycle time.
- Rungs 19, 20: These rungs are used to reset both retentive timers at the end of the shift, after the value of the accum registers have been recorded.

6.9 SEQUENTIAL FUNCTION CHARTS AND GRAFCET

We have described the process of creating a ladder diagram as an intuitive process. Like all computer programming, there is a good deal of insight required by the programmer in writing the code; five programmers addressing the same problem will often come up with five different solutions.

A fundamental problem in programming control systems is that the interaction of the software with inputs and outputs begins to get difficult to conceptualize as the size of the problem grows. What is required is a structured approach to specifying the control program. In this section we shall discuss an approach to organizing and documenting the control logic before developing a program. This approach is aimed at easing the software development conceptualization process and providing easy-to-read program documentation for those that have to maintain and revise the program. We shall refer to this approach as the graphical programming language (GPL) approach. The GPL is analogous to a flow chart in high-level programming. It is a structured representation of the program flow.

The genesis of GPL is fairly recent, going back to the late 1970s. In 1979, the French Association for Economical and Applied Cybernetics introduced a formalism called GRAFCET, which is a graphical language for specifying controller logic. GRAFCET employs a simple graphical representation of the states in a control program and the events that trigger the evolution of the program from one state to another. It has a powerful advantage over ladder logic in that understanding the control program is much easier, even for those not having a control automation background.

During the 1980s the use of GRAFCET grew rapidly in Europe. By the end of the 1980s it was clear that GRAFCET had a commercial future, and several companies developed GRAFCET compilers for programmable controllers. An international standard was adopted and released by the International Electrotechnical Commission as IEC 848. The term **sequential function chart** (SFC) was adopted to describe the new programming standard, although the IEC standard is very close to the original GRAFCET formalism. In this discussion we follow some of the documentation practices of GRAFCET and SFC, but we shall not adhere rigorously to all the documentation practices. Therefore, we shall sometimes use a generic term 'graphical programming' to refer to these methods.

A GRAFCET or SFC is defined by three basic elements: **steps, transitions** and **flow lines**. The symbols used are shown in Fig. 6.23. There are three kinds of steps. The **initial step** (Fig. 6.23a) is a double-sided box. When the control program is initiated, the processor starts executing the program at this step.

The **step** (Fig. 6.23b) is analogous to a rung of the ladder in which an action takes place. For example, an action may be to turn on a particular output. A step is shown as a single-sided box. Each action in the control program has its own step. The specific action is described in a rectangle at the right of the step.

SEQUENTIAL FUNCTION CHARTS AND GRAFCET 273

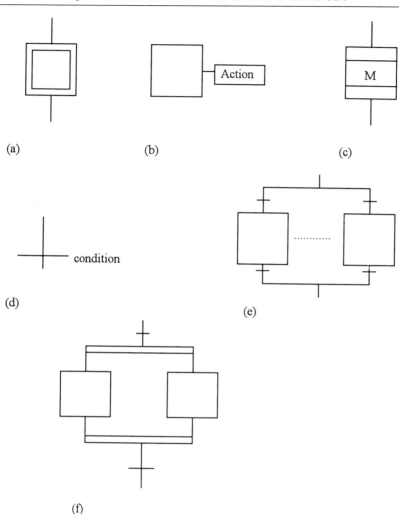

Fig. 6.23 Symbols for sequential function charts.

A **macrostep** (Fig. 6.23c) is a collection of steps. As programs become more complex, it is convenient to collect a portion of the program in a single step that describes the overall function of the individual steps. This is analogous to a subroutine which consists of several individual lines of program code. The macrostep is used to describe the overall function of a collection of steps in the program. The steps are the individual executable lines of code.

The symbol for a **transition** (Fig. 6.23d) is a bar. A transition is an event that is considered to have zero time duration. The transition allows program control to pass from its input steps to its output steps when the enabling condition of the transition is true. The enabling condition is written to the right of the transition. Transitions control the activation and deactivation of steps. When the processor evaluates the transition to be true, the step preceding the transition is disabled and the step following the transition becomes active.

The flow of the program logic may allow a transition to occur from one or more steps to one or more other steps. The **selection branch** (Fig. 6.23e) is the GRAFCET and SFC equivalent of a logical OR condition. The processor selects one of the steps to branch to depending on which transition goes true first. This structure is identified by a horizontal single line at the beginning and end of the structure.

The **simultaneous branch** (Fig. 6.23f) is the GRAFCET and SFC representation of a logical AND structure. When the processor enters this structure it executes all parallel steps simultaneously.

6.9.1 Graphical programming example

In order to illustrate the use of a graphical programming language, we shall apply the above concepts to the problem we have already studied using ladder logic, the drill press example. For convenience, we review the control steps required:

1. When power is turned on, enable the power-on lamp.
2. If the upper limit switch is open, raise the drill head until it trips the upper limit switch (homing) and enable the ready lamp. If the upper limit switch is closed, enable the ready lamp.
3. When the start cycle button is pressed, turn on the $-Z$ motor drive, the spindle motor, and clamp the workpiece.
4. When the lower limit switch is reached, turn off the $-Z$ drive and turn on the $+Z$ drive.
5. When the upper limit switch is reached, turn off the $+Z$ drive, the spindle motor and the clamp. Go to step 3.

Figure 6.24 is an implementation of the above requirements. The graphical program is labeled with notations that indicate the role of each step and transition. Conditions that enable transitions are written next to the transition and actions that result from a step are written in a box to the right of the step.

The program begins with the initial step, which is indicated by the double box. When power is brought to the machine, the program starts by executing the initial step, which is to enable the power on lamp. At that point step 1 is active.

The SFC then follows either one of two paths as given by the selection branch. If I4 is false, the left branch is selected. If I4 is true, the right branch is selected. The term **firing** is used with respect to the execution of a transition. A transition will fire when its input steps are active and its enabling condition occurs. The **enabling condition** is the condition written to the right of the

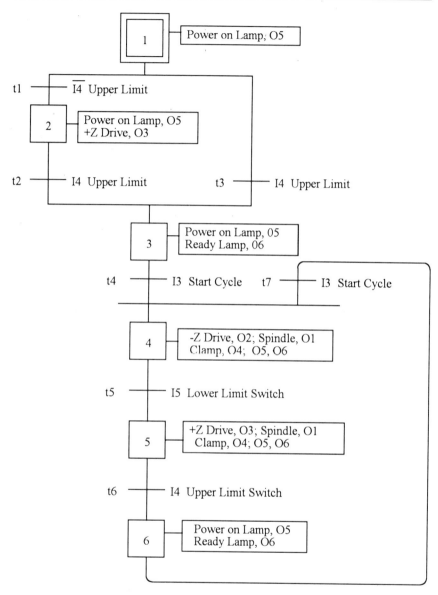

Fig. 6.24 Graphical program for drill press operation.

transition. Thus, transitions t1 and t3 both have their input step active when step 1 is enabled. The transition that fires first will depend on which enabling condition is true.

Steps 1 through 3 are those associated with the homing cycle of the machine. Transition t4 is the entry transition to the machining cycle. Steps 4, 5 and 6 are the machining cycle steps, which terminate at t7. Thus, once t4 fires, the program enters a machine cycle that continues to execute until power is cut off. Note the entrance conditions for step 4. Step 4 will be enabled if either t4 OR t7 fire. Firing transition t4 requires that step 3 is active and transition t7 requires step 6. This structure is known as a junction OR and is identical to the structure for t2 and t3. It is left to the reader to study the structure of the diagram in relation to the requirements of the problem as given above.

A comment should be made about the role of the action rectangle. In our illustrations in this chapter, the action rectangle of a step describes all of the outputs that are enabled during the execution of that step. Therefore, in step 4, there are five outputs that are on: $-Z$ drive, spindle, clamp, power on lamp and ready lamp. It will be shown in Chapter 7 that there are alternative ways in which to describe the action of a step.

Note the improvement in traceability of this program over ladder logic. The flowchart-like representation is easier to follow, even for those not specifically trained in control automation. The reader should compare Fig. 6.24 with Fig. 6.17 to gain some appreciation of the graphical programming approach.

It is also possible to represent the overall structure of the program using macrosteps. This is shown in Fig. 6.25. Here the distinctive cycles are shown as

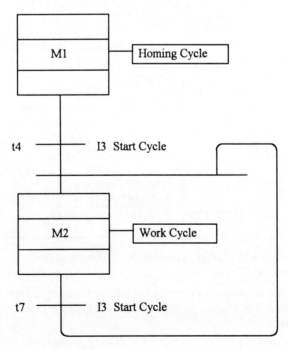

Fig. 6.25 Macrosteps for program of Fig. 6.24.

SEQUENTIAL FUNCTION CHARTS AND GRAFCET

well as the key transition that takes the program from one cycle to the other. Macrosteps give a higher-level view of the problem and each macrostep can be decomposed to reveal the lower-level execution of the program.

As mentioned earlier, companies have developed compilers that will give executable code from the graphical description shown in Fig. 6.24. These compilers result in boolean code that yields a ladder logic structure. We discuss the relationship between a GRAFCET or SFC graph and boolean equations in the next section.

6.9.2 Relationship between a graphical program and boolean equations

The purpose of this section is to illustrate how boolean logic can be created from a graphical programming representation of a control problem. The procedure follows the rules by which GRAFCET and SFC evolves. They evolve from one step to another by firing transitions. A transition fires when its

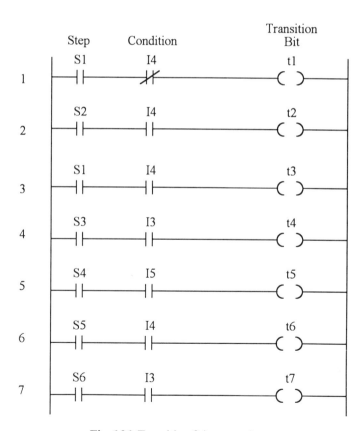

Fig. 6.26 Transition firing equations.

input steps are active and its enabling condition is true. When a transition fires, its input step is disabled and its output step is enabled. The above evolution is captured in two sets of boolean equations: the transition firing equations and the step activating/deactivating equations. We shall illustrate these equations as rungs in a ladder diagram.

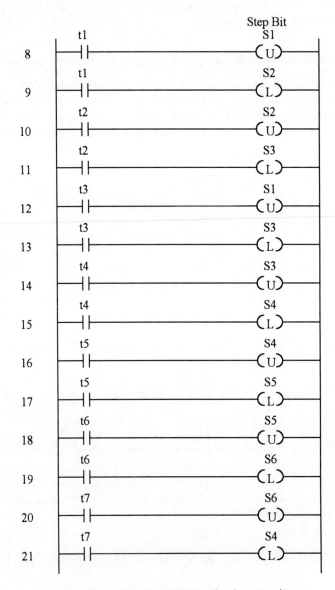

Fig. 6.27 Step activating and deactivating equations.

SEQUENTIAL FUNCTION CHARTS AND GRAFCET

Figure 6.26 shows the transition firing equations. In each case, the logical AND of the input step and the enabling condition fires the transition bit. Thus, t1 fires when bits S1 and I4 NOT are true. Here we indicate steps and transition bits by the prefixes S and t, respectively. In fact, these are information bits and would be assigned to a bit (B) file in the controller when the program is actually implemented.

When a transition is enabled, it is in its firing state. This means it must disable its input step and enable its output step. Rungs to satisfy this requirement are shown in Fig. 6.27. So, for example, when t1 fires, S1 is unlatched and S2 is latched.

Figures 6.26 and 6.27 account for the evolution of the program of Fig. 6.24 from step to step. Up until this point there have been no actions (outputs) enabled by the program. Referring to Fig. 6.24, we see that being in a step is coincident with taking the actions of that step. This is the logic used in Fig. 6.28, which shows the output enabling equations. For example, rung 32 shows that the power on lamp will be on during all steps of the program and rung 30 shows that the $+Z$ drive will be on during steps 2 and 5. Thus, when the bit representing a step is latched in the rungs of Fig. 6.27, the outputs of that step will be enabled in the rungs of Fig. 6.28.

Finally, it is necessary to initialize the ladder logic program. This is done by latching the initial step of the program on the first scan. In Fig. 6.29, step 1 is latched and all other steps are unlatched by normally closed B10. The last rung of the program latches B10, which has the effect of preventing rungs 22 to 27 from executing again until the machine is powered down and restarted.

Let us trace the first couple of scans of the program. Upon start-up, the controller scans the program, executing rungs 22 through 27. At that point, S1 is enabled and all other states are disabled. When the controller evaluates rung 32, it will be true and the power on lamp, O5, will go on. When rung 34 is executed, rungs 22 through 27 will be disabled.

On the second scan, either rung 1 or 3 will be true, firing either t1 or t3. Assuming t1 fires, the next rungs to be executed will be 8 and 9, which disable S1 and enable S2. Given this outcome, rung 30 and rung 32 are true, simultaneously enabling the $+Z$ drive and the power on lamp. The program continues to evolve in this manner. It is left to the reader to compare the remainder of the program with the requirements of Fig. 6.24.

The reader should by now have observed that there is more code required to program a problem using the suggested procedure as opposed to our earlier program directly using ladder logic. In Fig. 6.17, nine rungs were required; here we have used 34 rungs. This means that the scan time of the program should be increased in the SFC version. However, this may not always be true. One of the benefits of SFC is that is can be used to organize the program into modules with easy-to-identify events that move the program execution between modules. For example, in Fig. 6.25 we have shown that this program consists of two distinct modules and a transition, t4, which takes the program out of one

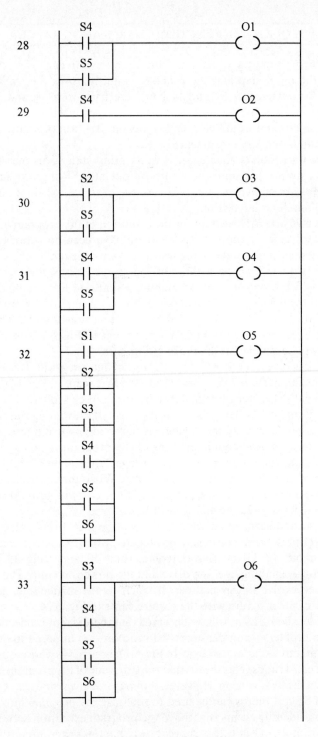

Fig. 6.28 Output enabling equations.

SEQUENTIAL FUNCTION CHARTS AND GRAFCET

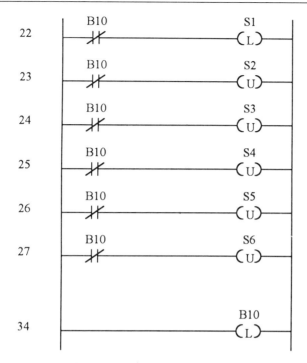

Fig. 6.29 Initializing the program.

module and into the other. Therefore, once t4 has fired, it is unnecessary from that point on to evaluate the rungs in macrostep M1. This has led to the practice of placing modules into different files of the controller's processor and just scanning the active file. When the exit transition occurs, the program jumps to the next macrostep file. By breaking a program down into modules, it is possible to maintain relatively fast scan times. Developing the structure of these modules and their transitions is enabled by the use of the graphical programming approach.

In general, the organization and modularity of the graphical programming approaches makes a program understandable and traceable. This is particularly important for maintaining a program and modifying it over time. Also, when the structure of the program is documented in SFC, the control logic is created according to a specific set of rules. This means that different individuals addressing the same problem structure will end up with the same ladder logic code. This kind of standardization is very desirable in software engineering.

In this section we have suggested a procedure for mapping a graphical program to a boolean control specification. This was done for illustrative purposes only in order to show the relationship that can exist between a graphical program specification and a ladder logic representation of the control program. However, the programming of complex control problems at a professional level and the generation of boolean code for programming should be done using an industry standard compiler. Such compilers closely adhere to the GRAFCET or SFC standard, include integrity checks of program logic, and have been tested over many complex problem conditions. The bibliography at the end of this chapter contains references to companies that are sources for these compilers.

6.9.3 A case study in graphical programming

To extend our illustration of the concepts of SFC, we introduce the control problem of Fig. 6.30. This is a mixer that has two cycles: a mixing cycle and a rinse cycle. When the operator presses the 'start product cycle' button, two materials are measured into hoppers by filling the hoppers until their upper limit switches (LS1 and LS3) are enabled. At that point the pumps are disabled

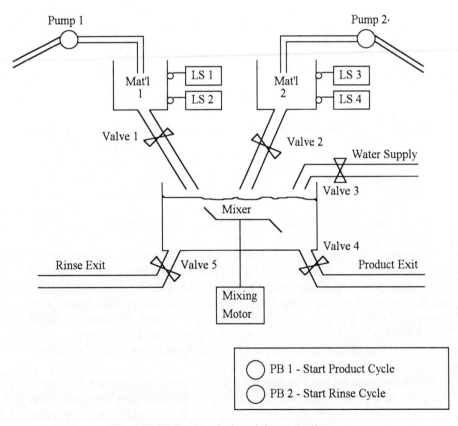

Fig. 6.30 PLC control of a mixing operation.

SEQUENTIAL FUNCTION CHARTS AND GRAFCET

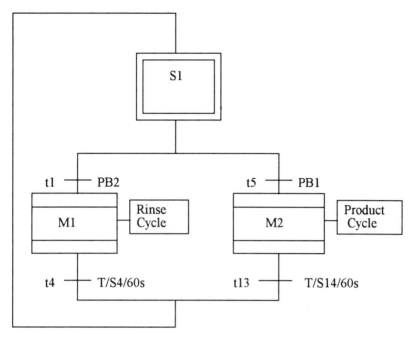

Fig. 6.31 Graphical program with macrosteps.

and the two hoppers simultaneously dump to the mixer by opening valves 1 and 2. The end of the dump is indicated by disabling the lower limit switches LS2 and LS4. When this is complete, the mixing motor mixes the materials for 10 minutes. When this is complete, valve 4 is opened for one minute, allowing the material to exit the mixer. The mixing cycle can begin again by pressing the 'start product cycle' button.

At the end of the shift and whenever the operator notices material caking on the mixer, the operator may put the machine through a rinse cycle. This is done by pressing PB2, opening valve 3 for one minute, followed by running the mixer for five minutes and flushing it out the rinse exit by opening valve 5 for one minute.

This case study is sufficiently complicated enough to make it useful to apply some of the structured techniques previously discussed. Using our general insight, it is possible to conceptualize the overall operation in a SFC format. This is done in Fig. 6.31, which illustrates how a programmer might produce a GRAFCET or SFC program using a top–down approach. The highest-level macrosteps are the two major cycles: the product cycle and the rinse cycle. Since these cycles are mutually exclusive, as indicated by the OR divergence, the program need only scan the macrostep that is active. Thus, when the initial step is active and PB2 is pressed, the program jumps to the file containing the

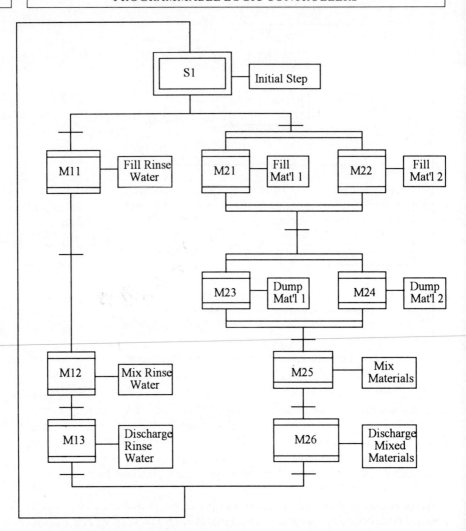

Fig. 6.32 Alternative macrostep structure for mixing operation.

rinse cycle. A further decomposition of the problem is shown in Fig. 6.32. Here we capture a greater level of detail of the problem structure.

Figure 6.33 represents a complete decomposition of the problem. Here, all the regular steps are shown with their entrance and exit conditions. Note the occurrence of parallel paths shown as simultaneous branches. In particular, S5 and S6 will both execute simultaneously when t5 fires. Note the use of dummy steps as in S7 and S8. Pump 1 must be turned off independently of pump 2. However, both pump 1 and pump 2 must be off (both hoppers full) before t8 may fire. The use of dummy steps has allowed these conditions to be satisfied.

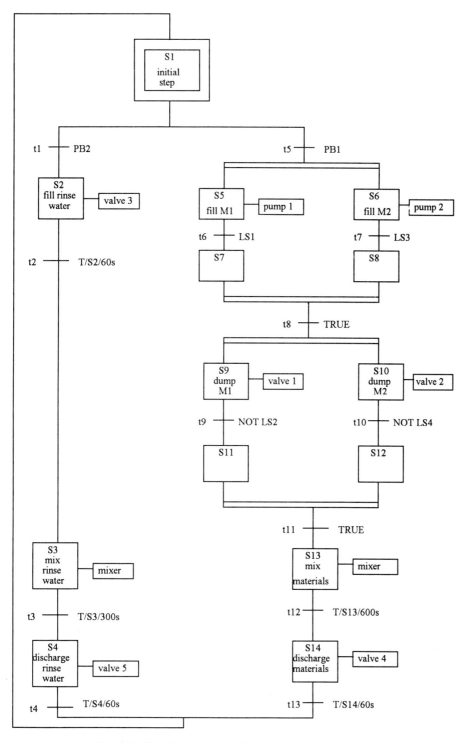

Fig. 6.33 Graphical program for mixing operation.

Finally, note that t8 and t11 are TRUE always and will fire as soon as their input steps are active.

There are several timers shown in Fig. 6.33. We use the convention of labeling a timer with the prefix T. This is followed by the step that enables the timer. Finally, the timer duration is given in seconds. The transition fires after its input step is active for the duration of the timer. In the ladder logic program that corresponds to the graphical program, the timer is associated with transition firing. Therefore, the timer would be programmed as part of the transition firing equations (as in Fig. 6.26). The step activating and deactivating equations (as in Fig. 6.27) would be governed by the timer DN bit. It is left as an exercise for the reader to incorporate timers and counters into the translation of a graphical program to ladder logic.

This case study illustrates a fairly involved problem utilizing many of the constructs of GRAFCET and SFC. The reader should closely examine Fig. 6.33 in relation to the requirements of Fig. 6.30. This problem will be used in exercises at the end of this chapter.

6.10 ADVANCED PLC PROGRAMMING

PLCs were first conceived as rather simple implementations of the computer, capable only of reading inputs, setting outputs, and setting bits for counting and timing. Indeed, this would replicate the functions of relay circuits, which the PLC was meant to replace. However, as technicians and engineers became more knowledgeable about computer technology, it became apparent that the PLC was not taking advantage of all the capabilities a microprocessor and memory could provide.

Two trends began to develop in the early 1980s. Manufacturers of microcomputers began to build machines that could be used on the factory floor and be directly interfaced to industrial processes. Other manufacturers developed I/O modules and interfaces that could perform digital and analog functions under the supervisory program of the microcomputer. The computer could be programmed in a high-level language and have all the capabilities of a general-purpose computer as well as a programmable controller. The major drawback was the requirement on the company to have people trained in a programming environment other than ladder logic.

Sensing these competitive changes, manufacturers of PLCs began to design more computer-like capabilities into their PLCs. Besides sequential logic, PLCs would be allowed to perform arithmetic and logical operations on bytes of data, communicate with other devices using RS232 or other protocol, and perform analog control functions. These advanced PLCs are more expensive and require additional programming knowledge. It is not our purpose here to give the reader a complete course in advanced PLC programming. However, we should describe some of the features of these machines so that the reader is aware of their capabilities.

ADVANCED PLC PROGRAMMING

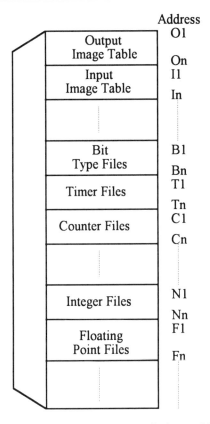

Fig. 6.34 Example memory map of advanced PLC.

6.10.1 Memory map and data structures

The place to begin is to examine the memory map of a typical advanced PLC. This is shown in Fig. 6.34. We already know that storage addresses are used for output, input, bits, timers and counters. To add arithmetic computations, addresses are added for integer and floating point arithmetic operations. The processor can be programmed to address these files using the syntax of the particular programmable controller.

It might be well to think about why the programmer might want to use these files. In Example 6.7, we wanted to keep track of the amount of time the drilling machine spent during changeover. Using a retentive timer, we collected that information throughout the day. However, suppose we want to collect the changeover time of each changeover. We could not do this unless we stored each changeover event. With integer files, this could be done by recording the timer value in an integer file at the end of each changeover. At a later point in time, the file could be read out of the PLC and the data analyzed.

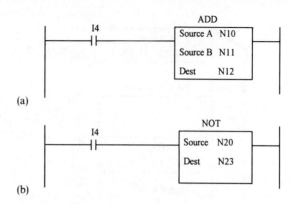

Fig. 6.35 Examples of arithmetic and logical operations in PLC ladder.

In the sections that follow we shall discuss a few of the arithmetic and logical operations of an advanced PLC, using a programming format that follows that of the Allen-Bradley Corporation.

6.10.2 Arithmetic and logical operations

Arithmetic and logical operations can be performed one word at a time. The format is shown in Fig. 6.35a. Here the rung is enabled when I4 is true. At that time the word in integer address N10 is added to the word in integer address N11 and the result is stored in integer address N12. When a logical operation is performed on two words, it is performed bitwise. Figure 6.35b shows a logical NOT command. Some typical operations available include:

Arithmetic	Logical
Add (ADD)	And (AND)
Subtract (SUB)	Or (OR)
Multiply (MUL)	Exclusive or (XOR)
Divide (DIV)	Not (NOT)
Square root (SQR)	

6.10.3 Move and clear instructions

It must be possible to move values between locations without change. This is done by the MOV instruction. Similarly, one clears a memory location of its contents using the CLR instruction. These instructions are illustrated in Fig. 6.36a and b. In 6.36a we move the accumulated value of counter C1 into integer address N11. In Fig. 6.36b we clear integer address N15.

ADVANCED PLC PROGRAMMING

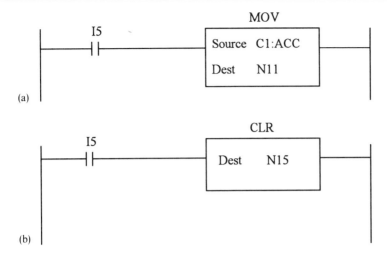

(a)
(b)

Fig. 6.36 Move and clear instructions in PLC ladder.

6.10.4 Conditional actions

In programming it is often desirable to perform an action conditional on some other event. Advanced PLCs have a comparison instruction set to enable this to be more easily accomplished. Typical comparison instructions are:

- Equal to (EQU)
- Not equal to (NEQ)
- Less than (LES)
- Greater than (GRT)
- Less than or equal to (LEQ)
- Greater than or equal to (GEQ)

In Fig. 6.37, input I3 enables a comparison between integer registers N5 and N6. If $N5 \geq N6$, the condition is true. If the condition is true the rung is enabled, enabling output O5.

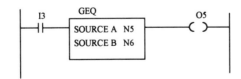

Fig. 6.37 Conditional Instruction in PLC Ladder

6.10.5 Continuous control

Advanced PLCs can be used in conjunction with analog devices for continuous control. This is typically done by assigning input and output addresses of the PLC to the analog device and communicating with the device by reading from and writing to those addresses. For example, consider an analog input module such as channels 3 and 4 shown in Fig. 6.2. This is simply an analog to digital converter with two input channels. The A/D converter reads the analog input signal, converts it and places a digital count in an integer register, where it is accessible to the processor. The data in the integer register can then be used in a control application. For example, a simple proportional controller is implemented by subtracting the digitized analog input from a setpoint value, multiplying this difference by a gain and providing the result as output to a D/A converter interfaced to the process.

6.11 SUMMARY

In this chapter we have described the architecture and use of the programmable logic controller. The PLC replaced relay circuits as the primary device for the control of machines and processes due to its flexibility and ease of programming. The programming language of PLCs, ladder logic, is a derivative of the symbology used by technicians in documenting relay circuits. More recently, GRAFCET and sequential function charts have been introduced as a documentation tool and a method for program standardization. In addition, PLCs have been augmented to include more computer-like capabilities that allow limited computation and data manipulation.

EXERCISES

1. Consider the PLC shown in Fig. 6.38. I want lamp 1 to go on when I1 is pressed, lamp 2 to go on when I2 is pressed, and neither lamp to go on when both I1 and I2 are pressed. Write the ladder logic program.

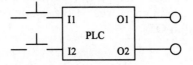

Fig. 6.38 PLC for Exercise 1.

2. A PLC controls the operation of two valves that fill and empty a tank, as shown in Fig. 6.39. When the tank is empty, float switches 1 and 2 are off. The controller opens valve 1. When the tank is full, float switches 1 and 2 are both enabled. The controller closes valve 1 and opens valve 2 until the tank drains. At that point the cycle begins again. Valve 1 is connected to O1 and valve 2 is connected to O2. Write the PLC program.

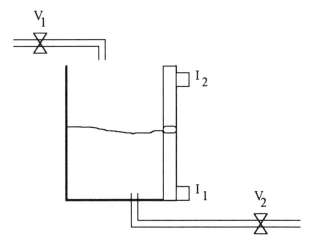

Fig. 6.39 Two valves filling and emptying a tank.

3. In section 6.5.2 there was a discussion of program design for failsafe conditions. Following the discussion of that section, redesign the program of Fig. 6.14 using normally closed limit switches.
4. In section 6.7 there was a discussion of the use of the master control relay (MCR) to disconnect power from PLC outputs. Following the discussion of that section, add MCR contacts to Fig. 6.17 as needed to maintain correct program status.
5. A PLC is operating two conveyors in tandem, as shown in Fig. 6.40. The conveyors are feeding the input to a machine. In order to prevent components from piling up on conveyor 1, the following control discipline is used:

   (a) Conveyor 2 cannot be started until conveyor 1 is started.
   (b) If conveyor 1 is stopped, conveyor 2 must stop.

   The inputs and outputs are:

   - I1 momentary contact push botton to start conveyor 1.
   - I2 momentary contact push button to stop conveyor 1.
   - I3 momentary contact push button to start conveyor 2.

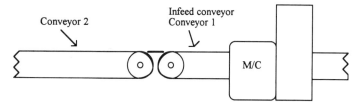

Fig. 6.40 Conveyors feeding a machine.

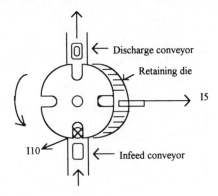

Fig. 6.41 Rotary indexing table.

- I4 momentary contact push button to stop conveyor 2.
- O1 conveyor 1 motor.
- O2 conveyor 2 motor.

Write the PLC program to run this conveyor line.
6. The machine being fed by the conveyor in Exercise 5 is a stamping machine. It consists of a four-position rotary table that captures the part at the infeed, indexes it 90 degrees, where it is stamped, and indexes it 90 degrees, where it is released to the discharge conveyor. This is shown in Fig. 6.41. The rotary index table, on output O3, indexes once each time it is turned on by the PLC, i.e. it is triggered by the leading edge of the PLC output. An index takes approximity one second. A proximity sensor at the infeed enables I10 when a part is present. A proximity sensor placed in the retaining die enables I5 of the PLC when a workpiece is in the stamping position. The PLC initiates a stamping cycle by turning on output O4 when it wants the stamping tool to descend. At the bottom of the stroke the stamping tool engages a lower limit switch, PLC input I6. The PLC then turns off O4 and the stamping tool ascends automatically. It takes approximately two seconds for the stamping tool to fully retract.

In order to enable continuous operation of the stamping press, the operator must press push button I11. The stamping press can be stopped by pressing an E-stop, which shuts down power to the machine and opens a MCR contact which is input I12 to the controller. Add these new requirements to the PLC program of Exercise 5.
7. In order to ensure worker safety, it is common practice to install safety shields at the infeed of machinery. This is normally done by mounting a sliding glass panel in front of the working area of the machine. When the safety shield is in its fully down position, it is depressing two push button contacts. The PLC is programmed so that it will not run the

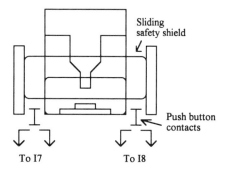

Fig. 6.42 Safety guards on a machine.

equipment when the safety shield is not fully down. This is to prevent accidental injury if the the operator must reach into the machine to clear a jam.

For the situation of Exercise 5, assume a safety shield has been installed as shown in Fig. 6.42. Write the appropriate PLC program.

8. For the drilling machine of Example 6.5 in the text, we wish to add a software stop button and a safety shield. The software stop button will allow the operator to interrupt a cycle by pressing the software stop button, I10. When this is done, the drill immediately goes to its home position and stops. The safety shield has two push button inputs, I11 and I12. The drill will not run unless I11 and I12 are enabled. If they are disabled when the machine is in operation, the machine will stop. Revise the ladder logic diagram of Fig. 6.17.

9. A packing and palletizing machine packs cans of soup into cartons, seals the cartons, and packs them on a pallet. There are 24 cans to a carton and 72 cartons to a pallet. Cans are counted by a photoelectric sensor. When 24 cans have passed the sensor, the PLC turns on the cartoning machine for five seconds. The machine automatically loads the 24 cans into a carton, seals it, and places it on the pallet. A proximity switch beneath the carton station indicates the presence of a carton, which is a requirement for the machine to go through the cartoning cycle. If no carton is present, the PLC turns on a red lamp to notify the operator that cartons should be loaded to the machine.

When 72 cartons have been loaded on the pallet, the PLC signals the machine to remove the pallet. At that point the packaging machine removes the full pallet, replacing it with an empty pallet. The PLC monitors a proximity switch at the pallet station. After signalling for a pallet removal, the switch should go false (pallet removed) and then go true (new pallet at station). If the signal is not returned in 20 seconds, the PLC turns on the red lamp. The cycle begins again. The inputs and outputs are shown in the following tables.

Input	Purpose
I1	A photoelectric sensor to count passing cans
I2	A proximity sensor at carton station
I3	A proximity switch at pallet station

Output	Purpose
O1	To turn on packing cycle
O2	To turn on carton removal cycle
O3	To turn on red lamp

Develop a ladder logic diagram.

10. A PLC operates the conveyor line shown in Fig. 6.43. There are two kinds of parts that come down the line: Hi and Lo. Lo triggers sensor S2 only. Hi triggers both S1 and S2. If the part is Lo, it is diverted using a gate. If the part is Hi, it goes straight through. S3 signals the clearing of Lo. The inputs and outputs to run the system are shown in the following tables.

Input	Purpose
I1	Push button to start the conveyor
I2	Push button to stop the conveyor
I3	Sensor S1
I4	Sensor S2
I5	Sensor S3

Output	Purpose
O1	Conveyor motor
O2	Diverter gate

Write a PLC program to operate the system and keep track of the number of Hi and Lo parts sent down the conveyor.

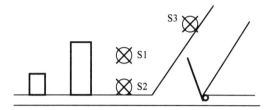

Fig. 6.43 Conveyor sorting parts.

11. A robot work cell is enclosed by a fence to prevent workers from entering the workcell and being injured by the moving robot arm. There is a gate to the enclosure through which a maintenance worker can enter. To gain entry, the maintenance worker swipes a card through a magnetic card reader for identification. If the card identifies the person as a maintenance worker, the reader outputs a signal to the PLC that controls the security gate and the gate lock is disabled. The following inputs and outputs apply:

Input	From	Message
I1	Card reader	0: nothing; 1: unlock gate
I2	NC limit switch	0: gate closed; 1: gate open
I3	NO pushbutton	0: nothing; 1: restart the robot

Output	To	Message
O1	Robot	0: power off; 1: power on
O2	Gate lock	0: lock gate; 1: unlock gate
O3	Alarm	0: alarm off; 1: alarm on

The sequence of operations is as follows:

(i) If a maintenance worker swipes an ID card through the card reader, the gate will be unlocked. The worker then has five seconds to open the gate. If the gate is not opened in five seconds, the gate will be locked again. If the gate is opened within five seconds, power to the robot will be disabled.

(ii) When the maintenance worker enters the enclosure, the worker must close the gate to prevent others from entering. If the gate is not closed within 30 seconds, an alarm will be sounded. The alarm is disabled by closing the gate.

(iii) When the maintenance worker leaves, the gate is closed behind him and the ID card is swiped through the card reader again. The PLC then locks the gate.
(iv) Power is returned to the robot by pressing the restart button.

   (a) Prepare a GRAFCET diagram of the operation of the system.
   (b) Prepare a ladder logic diagram.

12. Figure 6.44 shows one implementation of a counter in a graphical programming language. The counter is indicated as an action of a step. When the step is enabled, the counter C5 indexes by one. Using the graphical programming language, extend Fig. 6.24, the drill press problem, to include the counter functions of Fig. 6.20 and the timer functions of Fig. 6.22.

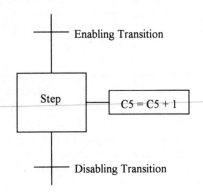

Fig. 6.44 Symbol for a counter.

13. For the problem of Exercise 12, create the ladder logic program from the graphical program following the rules of section 6.9.2.
14. A food company prepares a starch solution in a mixer as shown in Fig. 6.45. Input O0 is a push button that starts the process. When the process is initiated, valve V1 is opened and water is allowed to enter until LS2 is enabled. At that point valve 1 is closed and valve 2 is opened to add a starch slurry to the mix. When LS3 is enabled, valve 2 is closed. At this point the mixing motor and the heater are both turned on. Two minutes later, the controller begins to sample the temperature and viscosity of the process, while the mixer and heater continue to run. The temperature probe provides two inputs to the controller, I5 and I6. If I5 is true, the temperature of the solution is too high and the heater should be turned off. If I6 is true, the temperature of the solution is too low. When both I5 and I6 are false, the temperature of the solution is in the correct range. The viscosity probe has two inputs to the controller, I7 and I8. If I7 is true, the

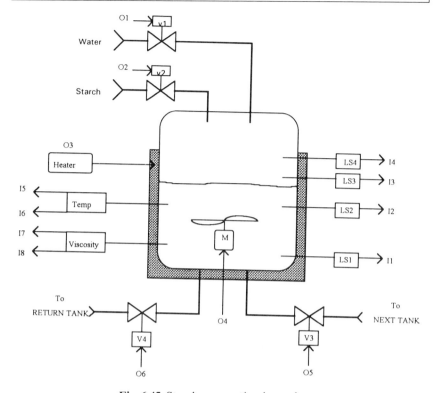

Fig. 6.45 Starch preparation in a mixer.

solution is too thick and water should be added. If I8 is true, the solution is too thin and starch should be added. If both I7 and I8 are false, the viscosity is within the acceptable range.

When the solution is in the acceptable range for both temperature and viscosity, the heater and mixer are shut off and the contents of the tank is dumped to the next tank by openning valve V3 until LS1 is false. If, while adding water or starch, LS4 is enabled, the input valve is closed and valve V4 is opened to reduce the contents of the mixer until LS3 becomes false. At that point, the process of testing for acceptable temperature and viscosity continues.

(a) Develop a GRAFCET for this process.
(b) From the GRAFCET, develop a ladder logic diagram.

15. Refer to the case study of section 6.9.3 and its graphical solution in Fig. 6.33. Using the methodology of section 6.9.2, develop the ladder logic program by the following segments: transition firing equations, step activating and deactivating equations, output enabling equations and initialization equations.

FURTHER READING

Allen-Bradley (1987) *PLC-5 Family Programmable Controllers Processor Manual*, Allen-Bradley Company, Inc., Milwaukee, Wisconsin.

Allen-Bradley (1993) *Advanced Programming Software Reference Manual*, Allen-Bradley Company, Inc., Milwaukee, Wisconsin.

Boucher, T. O. and Sung, P. (1994) Developing a class of sequential function charts from Petri nets, *Proceedings of the Fourth International Conference on Computer Integrated Manufacturing and Automation Technology*, IEEE Computer Society Press, Los Alamitos, California.

Baracos, P. (1992) *Grafcet Step by Step*, FAMIC Automation, Inc., Montreal, Canada.

David, R. and Alla, H. (1992) *Petri Nets and Grafcet*, Prentice Hall, Englewood Cliffs, New Jersey.

Lentz, K. W. (1994) *Design of Automatic Machinery*, 2nd edn, Chapman & Hall, London, England.

Lloyd, M. (1985) Graphical function chart programming for programmable controllers, *Control Engineering Magazine*, October, Cahners Publishing Company.

Pessen, D. W. (1989) *Industrial Automation*, John Wiley & Sons, New York.

Telemecanique (1987) *Grafcet Language*, Telemecanique, Inc., Westminster, Maryland.

Webb, J. W. and Reis, R. A. (1995) *Programmable Logic Controllers: Principles and Applications*, 3rd edn, Prentice Hall, Englewood Cliffs, New Jersey.

Supervisory control of manufacturing systems

7

7.1 INTRODUCTION

In Chapters 5 and 6 we examined the two main control models encountered in the design of unit operations. Control theory (Chapter 5) provides the basis for modeling and developing software for control of processes where process variables are changing continuously in real time. The PLC and the graphical programming language approach (Chapter 6) to modeling provides the basis for modeling and developing software for sequential control, where the states of the process change in discrete instances of time. In both of these chapters we limited our attention to unit operations, such as machines and batch process equipment.

It was pointed out in Chapter 1 that there is a hierarchy of control issues in manufacturing, from the machine or unit operation level up through the overall control of plant operations. This heirarchy is reproduced again below.

Level 4: Plant	Order processing Purchasing Aggregate production planning Accounting	
Level 3: Shop floor	Materials management Quality management Shop floor scheduling	
Level 2: Work cell/ production line	Materials handling Part sequencing Inspection/Statistical process control	
Level 1: Machine	CNC machine tools Robots Programmable controllers	

The **machine control level** is responsible for ensuring that the sequence of machine operations correspond to the planned sequence, or programmed steps. Typically, the sequence of operations is carried out as prescribed by the program resident in the machine controller and there are few or no decisions to be made. The material of Chapters 5 and 6 are appropriate to this level of control.

At the **production line or work cell level**, the objective is to supervise the interactions between a group of related machines or processes. This level of control is not concerned with the operation of the machine or process itself – that is the responsibility of the machine control level. Examples of control decisions at this level include control of materials handling among machines fabricating a component and the decision to extract out-of- specification components while they are being processed on a production line or within a manufacturing cell.

At the **factory floor control level**, decisions are made that affect groups of production lines or work cells. For example, several production lines or work cells may be serviced by the same materials handling system that brings raw materials from storage to production to be manufactured into finished product. Since this materials handling resource is shared among production lines and work cells, there must be a supervisory level of decision making that decides how to allocate this resource, particularly when conflict occurs, i.e. when it is required to service two lines at the same time. Other examples include the scheduling of whichever line or cell will fabricate a part when there is a choice, and the management of common facilities, such as storage areas for raw materials and finished product.

At the **plant level**, control decisions are less concerned with the daily operation of the factory and are more closely related to the business objectives of the firm. A typical plant-level control decision is **aggregate production planning**, which refers to the process of planning the use of the production capacity of the plant to meet customer demands over a period of months or a year. The output of this plan is a schedule of which products will be produced during each period of time going forward over the period of the plan.

The control hierarchy illustrated above starts from the top and works its way down. For example, the plant-level aggregate production plan sets the overall boundaries of which products will be produced and when. This provides a constraint on the shop floor level, which must then allocate the required production to machining cells and/or other production processes in the most effective manner. Once a specific mechining cell or set of production processes is allocated its production schedule for a specific day, it is the responsibility of the work cell/production line level to coordinate the manufacture of the product through the related machines and processes it requires. Finally, when the machine is assigned its role in partially fabricating the product, it is the responsibility of the machine-level controller to execute the correct steps of the fabrication process. This paradigm has led to modeling the manufacturing problem as a hierarchy of decisions, where the upper levels of the hierarchy place constraints on each succeeding lower level. The objective is to assign each control decision to the lowest possible level in the hierarchy. The complete integration of all of these levels of decision processes, supported by computer information systems, is often referred to as **computer-integrated manufacturing**.

In this chapter we focus on the intermediate levels between machine control and factory control. The term **supervisory control** best describes the control problem of the intermediate levels. Supervisory control refers to the need for overall control of a number of interacting machines or groups of machines and processes, usually referred to as a **manufacturing system**. This supervisory control may be at the work cell/production line level or it may be at the shop floor level.

Whereas the techniques for continuous and sequential control of machines and processes are well developed, the supervisory control of manufacturing systems is still an area of considerable research and many of the solutions implemented in industry are not generic. Instead, they are solutions that have been found to work for particular circumstances.

There are several topics to be covered in this chapter. First, we introduce some classifications of manufacturing system designs based on their organization of production. The **organization of production** refers to the manner in which machines, workers, and materials are related to each other in the steps of manufacturing the product. Next we discuss control architectures and the communication requirements of supervisory control. The communication network is the glue that brings the separate unit operations under the control of one supervisory controller. Finally, we shall address the software specification needs. This will be done by introducing a formal modeling approach by which the control logic can be specified, tested and programmed.

7.2 PRODUCTION ORGANIZATION AND MANUFACTURING SYSTEM DESIGN

The manner in which a manufacturing system is organized has often derived its form from the processing requirements of the product and the technologies available to provide those processing requirements. This is especially true of the process industries, such as petroleum, chemical, pharmacuetical and food industries.

The **continuous process design**, illustrated in Fig. 7.1, has several production steps, or processes, linked together to provide continuous inflow and outflow of processed materials. Petroleum refineries and most chemical plants are large-scale production facilities that use the continuous process design model. In general, these designs provide high production rates but are dedicated to the production of a narrow range of products.

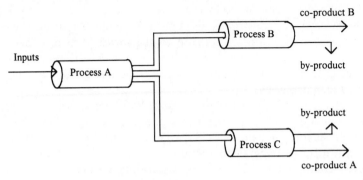

Fig. 7.1 Continuous process design.

The primary supervisory control problem in continuous processes is to maintain the setpoints of the process. This is more difficult than maintaining setpoints in a single loop controller, as was discussed in Chapter 5. These plants typically have multiple inputs and multiple outputs and the control of each variable is interactive with other variables. These interactive effects lead to the development of quite sophisticated control models which require in-depth treatment of the process thermodynamic and transport phenomena. Plants based on continuous process control were the first to use automatic control technology.

The **batch process design** is characterized by a sequence of several unit operations that accept and dispatch material in discrete batches as opposed to continuously. In general, the batch process facility tends to be more flexible in terms of the number of products it can produce, but the throughput rate is lower than that of a continuous operation.

Typical batch process designs include those of the food and pharmaceutical industry as well fine chemicals. Figure 7.2 shows the batch processing steps for a typical packaged food manufacturing plant. Materials are first processed in a kettle, where they are mixed and precooked. This is usually followed by a packaging or canning operation, where the mixed and processed material is filled into containers. Containers are then thermally processed in a retort (thermal sterilization process) in order to kill microbes that may reside in the sealed container. After sterilization, containers are labeled and packaged for shipment.

The supervisory control problem in the batch process design is to maintain the timing of events between successive operations and to monitor the quality of the product as it moves through the various operations. The term **recipe** or **formula** is used to describe the combination of materials to be used in a product and the time, temperature and other processing parameters of the unit operations that are used to manufacture the product. It is a supervisory

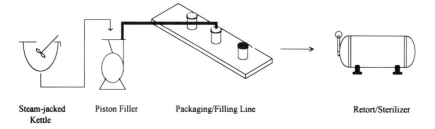

Fig. 7.2 Batch processes in packaged food industry.

control problem at the shop floor level to ensure that the correct recipes or formulas are downloaded for the planned production on a particular day.

In the fabrication and assembly of mechanical products, the traditional manufacturing system design has been the **functional** or **job shop** organization of production. This is shown in Fig. 7.3. In this design, machines that perform the same manufacturing processes are grouped together within the same departments. Hence, there is a department for lathes, which perform the removal of metal on cylindrical workpieces, a department for milling machines, which perform metal removal on prismatic parts, and so forth. Workpieces are routed among departments in a prescribed sequence, called a **process plan**. They emerge from that sequence as a finished component. From there they may be assembled with other components to make a finished product.

This manufacturing system design is typically used for the manufacture of components in small batches. It is a very unfavorable organization of produc-

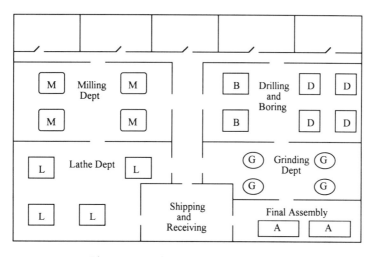

Fig. 7.3 Functional or job shop design.

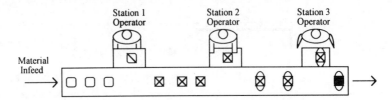

Fig. 7.4 Flow line design.

tion in the sense that it is difficult to manage the movement of batches of components through the plant in a timely fashion and to ensure an even balance of work in the various departments without investing in large quantities of work-in-process inventory. In addition, it does not lend itself very well to factory automation, except at the machine level. However, it is the most flexible of mechanical parts manufacturing systems in the sense that all the processes are available to work on any product.

The **production flow line**, illustrated in Fig. 7.4, is a method of organizing production in the mechanical industries such that individual fabrication operations are organized along the steps of manufacture of the product. The automotive assembly line is an instance of flow line production as is the manufacture of metal containers. These designs tend to be composed of closely coupled unit operations linked by conveyors. They are dedicated to a narrow range of products and are relatively high speed. A high degree of automation along the production line is not unusual.

The problem of supervisory control in flow lines is typical of that in tightly coupled systems. Supervisory control problems include the on/off control of conveyors that move product between operations, the monitoring of buffer capacities between operations, and the monitoring of jams and other stoppages of the line. Although this is the easiest of system designs to control in the metal fabrication and assembly industries, its high dedication of equipment to a narrow set of products does not lend itself to many applications where products are produced in small volumes.

A form of production organization in the mechanical parts industries that is intermediate between flow line and job shop is called **cellular manufacturing**. Cellular manufacturing takes advantage of some of the favorable aspects of job shop production and flow line production. It is based on the group technology manufacturing philosophy. **Group technology** identifies components of the product mix that are similar in design and that require roughly the same kinds of manufacturing operations. Hence, components requiring primarily external turning operations, such as shafts, are collected in one group, while components requiring surface grinding operations and drilling operations, such as plates, are assigned to a different group. These groups become the basis on which production engineers can reorganize a traditional job shop into a plant design with machining cells in which machines are arranged such

PRODUCTION ORGANIZATION AND SYSTEM DESIGN

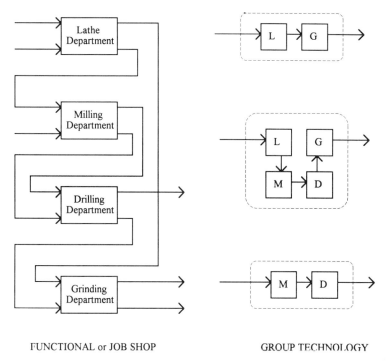

Fig. 7.5 Product flow under functional and group technology plant designs.

that each machining cell can complete the fabrication of one or a few groups of parts. Despite the fact that each individual component may have a small annual demand, when components are grouped together they add up to a large enough annual production quantity to utilize the capacity of the dedicated manufacturing cell.

Figure 7.5 illustrates the difference in the way the product is routed among operations as between the job shop design and cellular manufacturing. Through the simplification of routings, the cellular design eliminates the difficult routing and control problems that exist in a job shop. In addition, there is a dramatic reduction in setup times on each machine because the fixtures and workholding devices for each machine are redesigned to accommodate the components in the group without major changeover. With respect to automation, the cellular design makes it economic to load and unload the machining operations automatically, e.g. using robots.

Figure 7.6 shows just such an automated cell. Here a robot is attending several machining operations. Incoming parts are placed in an input buffer and finished parts are put into an output buffer. Parts that have undergone some machining operations are in a work-in-process buffer. The program to control the fabrication steps performed on the workpiece by each machine

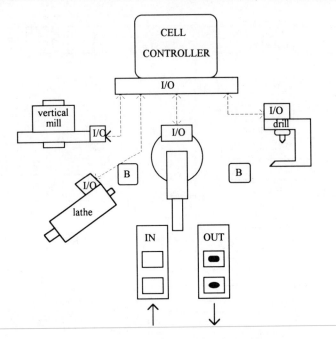

Fig. 7.6 Automated machining cell design.

resides in the controller of the machine. The program that determines how to load and unload components to and from machines resides in the robot controller. The program that provides overall supervision of the cell resides in the cell controller.

The supervisory control problems at the cell level can be of several kinds. They certainly include directing the robot when to load and unload machines as required and directing each machine to begin to execute its machining process at the appropriate time. It may also include monitoring the state of each operation and notifying a higher level in the case of suspected breakdown in the cell. If more than one component is being manufactured at a time, it is necessary to identify the incoming component and download the appropriate program to the machine tool before directing the machine to begin its operation.

The group technology philosophy has been extended to large scale manufacturing systems composed of machines and cells linked together by automated materials handling. The design is usually referred to as a **flexible manufacturing system**, or **FMS**. An example is shown in Fig. 7.7. Here we show manufacturing cells that are multipurpose machining centers serviced by a loading/unloading machine, sometimes a robot. These machining centers are capable of performing more than one machining operation on a single machine. They are linked together by an automated transport system. This

PRODUCTION ORGANIZATION AND SYSTEM DESIGN

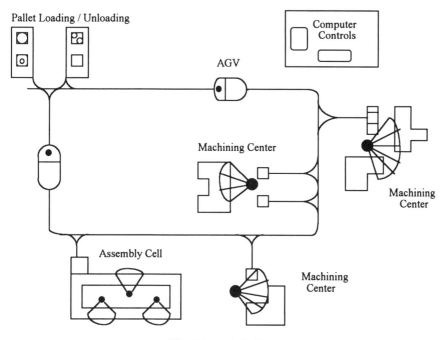

Fig. 7.7 FMS design.

may be an **automated guided vehicle** (AGV) that is programmed to follow a fixed path and carry components to be manufactured between cells, or it may be a pallet transport system that moves parts along an asynchronous conveyor on fixtures, called **pallets**. These pallets are of standardized sizes and they can be directly loaded onto the multipurpose machine tool when they arrive at the cell. There is no additional work required to fixture the workpiece at the machine. The setting of the workpieces on pallets is done by workers in a loading/unloading area. Hence, once a workpiece is fixtured on a pallet in the loading area, it stays on the pallet through its machining operations and transport back to the unloading area. This capability reduces machine setup time.

There is a hierarchy of supervisory control problems associated with the operation of a FMS. Assuming that a schedule of parts to be produced for the day has been determined and downloaded to the FMS controller, workpieces are shipped from raw material inventory into the FMS to be machined. The FMS supervisory controller calls for the fixturing of the workpiece when its scheduled time in the sequence occurs. When a workpiece is fixtured, the FMS supervisory controller must evaluate the machining requirements of the component, stored in the supervisory controller's data base, and determine the status of machines that can perform those operations as well as the availability of the required tooling at the machining centers. If there is only one machining cell that can perform the operation, the pallet is sent to the queue in front of that cell. If a choice of cell exists, the FMS supervisory controller evaluates the workload already assigned to each cell and selects a cell based on available machining time and the efficiency of using that machining cell for the particular component.

There are also some other resource sharing considerations that are managed by the FMS supervisory controller. In introducing workpieces into the system, the controller must consider the availability of the fixtures appropriate to mounting the workpiece. When directing the transport of fixtured workpieces among stations, the controller must consider the availability of a transporter and, if more than one route exists between stations, which route to take.

Regardless of whether the manufacturing system is designed for batch operations, or as stand-alone machining cells, or as a FMS, supervisory control of interactions among machines and cells requires communication links among the entities of the system. This requirement is usually addressed by local area networks and a communication hierarchy is part of the overall automation design. The next section discusses the applicaion of local area networks at the shop floor level.

7.3 COMMUNICATION ARCHITECTURE AND LOCAL AREA NETWORKS

Perhaps the most troubling problem in shop floor automation is that of interconnecting machines and processes. Historically, manufacturers of different equipment have implemented communication capabilities on an individual basis, without regard to the protocols adopted by other equipment vendors. Even the RS232-C standard, discussed in Chapter 2, is not uniformly implemented and it is often necessary for engineers or computer scientists to write their own device drivers when implementing communication between machines.

When manufacturers began considering the integration of unit operations, such as robots and machine tools, into coordinated production systems, it became obvious that point-to-point serial communication using RS232 was not a good solution to the connectivity problem. What was needed was a common medium over which the controllers of related machines could communicate with each other and with higher-level computers. Thus, the local area network concept was adopted by vendors of programmable controllers and by manufacturers of computers for shop floor data collection.

A **local area network** (LAN) is a communication network that is implemented over a limited area and is usually owned by one organization. It is a common medium that allows several machines to be connected and, as long

COMMUNICATION ARCHITECTURE AND LANs

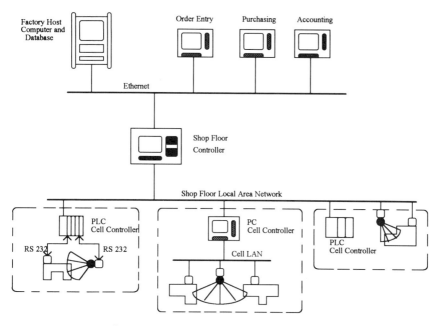

Fig. 7.8 Typical communication architecture.

as each machine uses the protocol conventions of the LAN, communication can take place at high data rates.

Figure 7.8 shows an example of a communication architecture on a factory floor. At the bottom of the architecture, cell controllers are wired to individual equipment controllers. The cell on the left-hand side is using RS232 to connect machine controllers to the cell controller. Thus, the RS232 protocol is used to communicate between devices. In the cell at the center, a cell LAN has been implemented. Here a computer is used as the cell controller and it communicates with the controllers of the individual machines over a common medium. This configuration is possible when the computer and individual controllers are all capable of using the protocol of the LAN. In the cell on the right, the cell controller and the equipment controllers are all capable of using the protocol of the shop floor LAN. Therefore, they communicate with each other over that medium.

The reader should appreciate that the architecture requiring the least additional complexity is that on the right of Fig. 7.8. There is no additional wiring; all machines communicate over the shop floor LAN. The reason for considering other configurations is the inability of all equipment controllers to use the shop floor LAN protocol. In the worst case, it is usually possible to use the RS232 or similar standard to set up point-to-point communication, as shown in the cell at the left. There are many different possible communication architectures for connecting shop floor equipment; Fig. 7.8 simply illustrates some examples.

There are a variety of LANs and protocols that exist in the manufacturing automation arena. Although choice is usually a good thing, this has created some practical problems when trying to find equipment controllers, the protocols of which are compatible. Indeed, this is the major issue in designing the factory communication architecture because it constrains the choice of production equipment that can be purchased without rewriting their communication protocols. This has led to a great deal of effort on the part of user groups to encourage LAN standards to be developed and adopted by equipment vendors.

There are several features that distinguish LANs. Among the most general features are the network topology, the transmission medium and the network access protocol. These will be described in the following three sections. Beyond these general features, there are specific differences in the encoding of data and the error checking and other facilities provided by the LAN. We will discuss these differences in the context of a reference model developed by the International Standards Organization and we will describe an example of a LAN and how it relates to this standard.

7.3.1 Network topologies

The **network topology** refers to the physical layout used to connect machines together on the network. The three topologies most often used are the star, ring and bus.

The **star** topology is shown in Fig. 7.9a. It consists of a central hub, which can be a computer, to which nodes are directly connected. The hub, or central computer, serves the function of a switch, providing point-to-point connection between nodes that wish to communicate with each other. Under this topology, the central computer can control communication, determining which machines will have priority on the network. Because all information flows through the hub, a record can be kept centrally of all the past activity on the network. A major drawback of this topology is the vulnerability of the network to failure of the hub; if the central computer goes down, all the nodes are affected.

The **ring** topology is shown in Fig. 7.9b. It consists of a continuous trunkline with nodes tapped off the trunk. There is no central or hierarchical control of the network. Each station has a unique address and station-to-station communication is accomplished by an addressing scheme. A message initiated by a station proceeds from one station to the next in a sequential fashion. Each station handles the message and, if it is not the intended recipient, passes it on to the next station. When the correct recipient gets the message, it copies it and sends an acknowledgement back to the address of

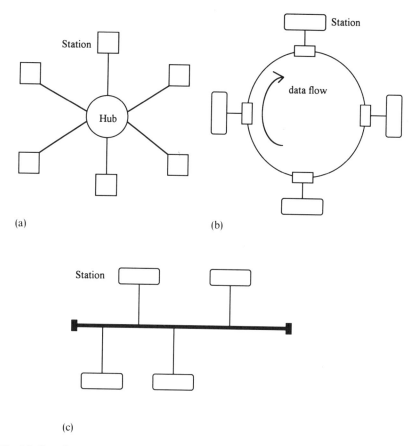

Fig. 7.9 Local area network topologies. (a) Star topology; (b) ring topology; (c) bus topology.

the sender. IBM promoted the ring topology, which eventually became part of the IEEE 802.5 standard.

The **bus** topology is the most common topology in shop floor manufacturing applications. It is shown in Fig. 7.9c. This topology uses a trunkline with terminal points; stations are connected along the trunkline. Unlike the ring, where communication is passed from station to station sequentially, the bus is a broadcast system. When a station sends out a message, it is heard by all the other stations on the LAN. However, the only station that will copy the message is the one to which it is addressed. This flexibility in transmission does create some control problems which we will discuss shortly. This topology has been adopted in two important network standards, IEEE 802.3 (Ethernet) and IEEE 802.4 (Manufacturing Automation Protocol or MAP). It is also used in the proprietary networks of the major programmable controller manufacturers, Allen-Bradley and Seimens.

7.3.2 Transmission medium

The **transmission medium** is the wiring over which the information physically flows. The most common medium are twisted wire pairs and coaxial cable.

When a pair of insolated wires are twisted together to improve noise immunity, they are called a **twisted wire pair**. For each pair, one line is ground and the other line is a data line. Figure 7.10a shows twisted wire pairs in a shielded cable. At a minimum a LAN requires two data lines, one for transmission and one for reception. Twisted wire pairs are very economical but have slower data rates than other mediums.

A **coaxial cable** consists of a center copper conductor surrounded by an insulator, which is itself surrounded by a mesh conductor and an outer insulator. This is shown in Fig. 7.10b. This cable provides very good noise

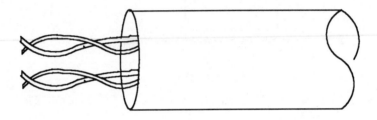

(a)

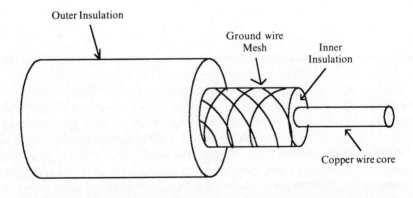

(b)

Fig. 7.10 Transmission medium. (a) Twisted wire pairs; (b) coaxial cable.

COMMUNICATION ARCHITECTURE AND LANs

immunity and, more importantly, can provide data transmission rates up to 50 Mbit/s (megabits per second). This compares to about 1 Mbit/s for twisted wire pairs. Both the Ethernet and MAP standards are based on the use of coaxial cable.

Coaxial cable can be further divided into baseband and broadband. **Baseband** coaxial cable carries one message at a time. Transmission can be accomplished as either analog or digital. The transmission speed is limited to 10 Mbit/s. Because it was part of the original Ethernet standard, it is probably the most common cabling used in local area networks today. Recently, manufacturers of Ethernet cards have introduced a twisted wire pair with a jack similar to telephone wire, called 10 baseT. It has two twisted wire pairs, one for transmission and one for reception.

Broadband coaxial cable allows several messages to be sent simultaneously and can carry multiple transmission types, such as voice, data and video. This is accomplished by transmitting at different frequencies. In order to do this and guarantee the quality of the transmission, a broadband system requires additional components and is more expensive to install. However, their data rates are in the range of 50 Mbit/s. Broadband transmission is analog in nature.

7.3.3 Network access

One of the general features that discriminate between types of networks is the manner in which transmission is controlled on the network i.e. the discipline by which stations gain access to transmitting on the network. The first commercially successful local area network was **Ethernet**. It is based on a bus topology and a baseband coaxial cable transmission medium. The technique used to govern network access is called **carrier sense multiple access with collision detection** (CSMA/CD). In this method, a station transmits a message anytime it wants. The transmission includes the address of the intended receiving station. Because it is a bus topology, the transmission is a broadcast message. All nodes on the network monitor the network for the transmission of messages. If a node wants to transmit a message and the network is clear of other transmissions, it sends its message. As long as only one station transmits at a given point in time, this discipline will succeed in providing a means of regulating transmission among nodes.

On the other hand, it is also possible that two or more stations will find the medium is clear and attempt to transmit at the same time. In this circumstance a collision of data will occur and the transmission will be corrupted. For that reason, the transmitting station monitors the signals on the line after sending its message. If the transmitted and monitored signals are not identical, a collision is assumed to have occurred and the transmitting station begins a recovery routine. In the recovery routine, transmitting stations each cease to transmit for a random interval of time and then try to retransmit the message.

The randomly determined interval of time is independently determined by each transmitting station. In all likelihood, the independently determined random intervals of each transmitter will avoid further collision and result in a successful retransmission.

There is one aspect of CSMA/CD that is disturbing in manufacturing automation applications. Since multiple collisions of data are possible, there is no guaranteed upper bound on the time it will take to deliver a message from one node to another. Since time is an important concern in automation, especially real time applications, this is one of the weaknesses of CSMA/CD in the manufacturing environment. If traffic on the transmission medium is heavy, this can be a serious limitation.

The other widely used method of controlling access on the transmission medium is through the use of a **control token**. This has been implemented as part of the IEEE 802.5 standard (Token Ring) and the IEEE 802.4 standard (MAP). Under this discipline, only one station is allowed to transmit at any point in time. This is accomplished by passing a token (control signal) from station to station. When a particular station gains the token, it may transmit a message or, if there is no message to transmit, it will pass the token to another station. In a token ring, the token is passed sequentially around the ring. In a bus topology, each station is assigned a successor station address to which to pass the token. Using a control token avoids the problem of data collision in transmission. It also places an upper bound on the amount of time a node will have to wait before it can transmit. For these reasons the control token discipline, or a variation of it, has been used widely in time-critical applications where real time control is involved.

7.3.4 The OSI reference model

In an effort to encourage some standardization in the design of communication networks, the **International Standards Organization** (ISO) has established a **reference model**, which is a description of how networks should be designed in terms of layers of responsibility. This model has become known as the **open system interconnect** (OSI) model. The purpose of the reference model is to divide the communication problem into a series of subproblems, or layers. There are seven layers in the OSI model: physical, data link, network, transport, session, presentation and application. Figure 7.11 shows two stations on a network and the relationship of the seven layers to each other. The lowest layer is the physical layer, which includes the transmission medium. The highest layer is the application layer, which interacts with user programs. Each layer in between has a specific function which will be described shortly.

The OSI model brings a divide-and-conquer strategy to the problem of designing a communication network. Each layer represents an independent subproblem that can be solved within the layer. Each layer interacts with the layer above and the layer below. Incoming messages at a station enter via the

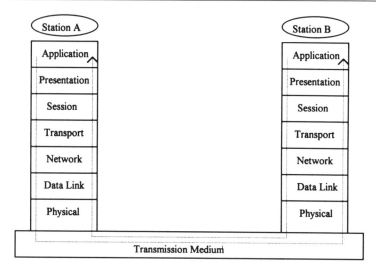

Fig. 7.11 Seven layer OSI model.

physical layer and are passed up the stack, with each layer providing a unique service to the message; outgoing messages travel down the stack. A network is designed so that the services provided at each level are the same for each station on the network.

The **physical layer** defines the rules governing the actual transmission of data over the network. There are several items that are governed at this level, e.g. the voltage levels to be used and the data rates for transmission are problems to be resolved by the physical layer. The hardware that is used to connect the node to the transmission medium and the type of transmission medium (twisted-wire pair, coaxial cable) are also part of the specification of the physical layer. The physical layer must also define whether transmission will be half duplex or full duplex. In half duplex, a message is sent and, afterwards, a message is received. Transmission and reception cannot go on simultaneously. Full duplex transmission allows transmission and reception to go on simultaneously. Although bits are transmitted using voltage levels representing 1s and 0s, there is no meaning given to the transmission at the physical layer.

The message unit on a LAN is called a **packet**. The packet includes the data to be transmitted along with other information, such as the address of the intended recipient node. Figure 7.12 shows the structure of an Ethernet packet based on the IEEE 802.3 standard. It begins with a preamble, which is used for synchronization, much like the start bits are used for synchronization in RS232 serial communication. This is followed by the address of the receiving node and the sending node. The length field (L) designates the number of data bytes being transmitted. The data field is the actual message. This is followed

Preamble	Destination address	Source address	L	Data	Frame check

Fig. 7.12 Structure of an Ethernet packet.

by a frame check sequence that is used to check the validity of the message received.

When the physical layer receives the packet, it passes it up to the **data link layer**, which provides certain integrity checks on the packet. For example, the data link layer determines that the message is for its station. It also checks for errors in the transmission based on an agreed-upon procedure using the frame check sequence. If the transmission is corrupted, it must return a response to the sender so that it can retransmit. In general, the purpose of the data link layer is to ensure that only valid messages are passed up to the next layer.

The **network layer** is used when there are two or more local area networks interconnected to each other. This situation is shown in Fig. 7.13. Network A and network B are two local area networks. If a message is to be sent from a node on network A to a node on network B, it must go through a device called a **router**. The router is a device that connects two LANs and is addressable from stations on each LAN. When a message sent by a station on one LAN is addressed to a station on another LAN, it is the responsibility of

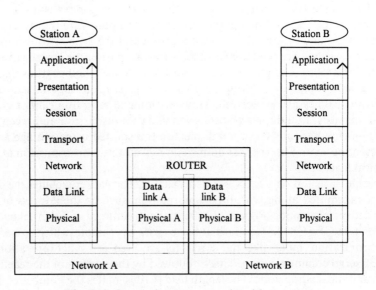

Fig. 7.13 Connecting two networks with a router.

the network layer to add information to the packet that gives the message its proper routing. Hence, the transmitted packet will contain a network header that will direct it to the correct local area network and a station header that will identify the station on the network. When the router receives the packet, it will delete the network header and pass the packet on to the correct LAN. Note that it is assumed in Fig. 7.13 that the two LANs are employing the same network protocols at layer 4 and above. Where all stations exist on the same LAN, the network layer is a null layer.

The **transport layer** is concerned with providing reliable end-to-end communication between stations. When a message is to be transmitted, a decision must be made to divide it up into packets. The transport layer is responsible for assembling the packets from the data passed down from the session layer. It is responsible for opening and maintaining a connection between end systems and ensuring that the packets arrive in the order they are sent. The transport layer is necessary where there is more than one LAN on the network and packets are being routed among LANs. With such inter-LAN routing, there is always the possibility that part of a message sent in an earlier packet will arrive after part of the same message sent in a later packet. Also, stations on different LANs may be operating at different speeds. The transport layer is responsible for slowing down transmission rates from a faster station if the receiving station cannot keep up with the transmission rate. If all stations are on the same LAN, the order in which packets are sent will be the order in which they arrive at their destination and the transmission and reception rates of the stations will be the same. This makes the transport layer unnecessary when the network is composed of only one LAN.

The **session layer** is concerned with the needs of particular users. For example, the session layer may verify the password of a particular user logging on the network. It may control the flow of data for the user by grouping the data into specified groups before presenting it to the user. It may also handle an orderly recovery from failure if the system crashes and it may handle the orderly termination of a user session.

Data can be encoded in a variety of forms for transmission. In Chapter 2 we examined ASCII code. There also exist other encoding schemes, such as Extended Binary Coded Decimal Interchange Code (EBCDIC). One of the responsibilities of the **presentation layer** is protocol conversion between computers using different encoding. Data bits entering this layer must be **presented** to the application layer in the code that is being used at that level. The presentation layer also has responsibility for data encryption when messages are being sent in a secure manner, as in the case of corporate financial information being sent across public networks.

The **application layer** is user specific. Typical applications are file transfers, database management and electronic mail. We shall shortly describe an application layer specific to communication with shop floor and machine controllers.

It is important to reiterate that the OSI model is a reference, or guide, to the development of communication networks. Specific implementations of networks are going to use different concepts from the OSI reference model. If the network architecture is simple, as it is in most manufacturing LANs, it will not be necessary to employ all the layers of the reference model. For example, when all stations are on the same LAN and use the same application protocols, there is no need for network, transport, session or presentation layers. These layers are more prevalent in wide area networks, where routing between nodes is more complex and different forms of data encryption are used. Section 7.3.6 describes in detail a widely used manufacturing local area network, the Allen-Bradley Data Highway communication network. This LAN employs three layers: physical, data link and application. Through a detailed explanation of how each layer executes its responsibility, we shall give some specific meaning to the previous description.

7.3.5 Routers, bridges and gateways

The function of a router was described in the last section. Basically, a router is necessary when stations that exist on two different LANs, having different physical and data link layers, are to communicate with each other. The router decodes the destination information provided by the network layer of the transmitting station and routes the packet to the network on which the destination station resides. This was illustrated in Fig. 7.13. In order to use a router to transfer data packets between networks, it is necessary that the

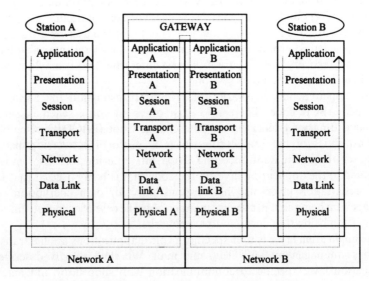

Fig. 7.14 Connecting two networks with a gateway.

transmitting and receiving network use the same protocols at layers 4 and above.

This raises the question of what to do when applications on different networks do not use the same protocols at layer 4 and above. In this case, a gateway is required for different applications to communicate with each other. The **gateway** provides a means for two dissimilar networks to communicate by performing a conversion at the application layer. Figure 7.14 illustrates the position of the gateway in the seven-layer OSI model. Gateways are available through manufactures of LANs as well as communication devices for connection between different networking protocols. There are also third party vendors that sell computer software that can transmit and receive using the protocols of different LANs. By connecting the transmission medium of these different LANs at the ports of a single computer, that computer can serve as the gateway between LANs. Gateway software sits on top of the application layer in the OSI model. With respect to manufacturing and shop floor applications, a gateway would be used if local area networks of different controller manufacturers were being used and it were necessary to communicate between stations on these different LANs.

Another common situation that can exist in manufacturing is that of connecting two LANs that have identical protocols, i.e. identical networks. Due to limitations on the number of nodes allowed on a single LAN, when a decision is made to expand automation applications within the factory, it may be necessary to install another LAN. If the LANs are identical, they may

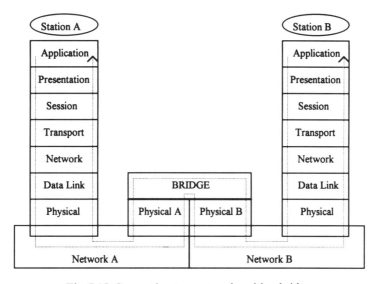

Fig. 7.15 Connecting two networks with a bridge.

be linked using a **bridge**. Because the two LANs use identical protocols, the bridge simply relays a transmission initiated on one LAN to another LAN on the network if the destination station number is on the LAN for which it is the bridge. Unlike a router, which makes modifications to the format of the packet due to differences between networks, a bridge does not change the packet. As shown in Fig. 7.15, the bridge just deals with the redirection of the binary string based on the contents of the destination frame.

7.3.6 Case study: the Allen-Bradley data highway communication network

We have chosen the AB Data Highway communication network as an example because of its wide use in manufacturing. There are different versions of Data Highway. Here we will discuss the Data Highway Plus™ network which supports the Allen-Bradley series 5 controllers and the Data Highway 485 network which supports the smaller 500 series controllers.

Figure 7.16 shows the layers of the Data Highway network. User-written application programs interface to an application layer which provides common application routines that are specific to the use of programmable controllers. The data link layer sets up the link protocol and the physical layer transmits the bits. The Data Highway Plus network allows peer-to-peer communication among 64 stations and the Data Highway 485 network accommodates 32 stations.

The physical layer for Data Highway specifies a bus topology and a token passing algorithm for access control. The transmission medium is shielded twin-axial cable or shielded twisted-wire pairs. The Data Highway Plus network has data transfer rates of 57.6 kbit/s and the Data Highway 485 network transmits at 19.2 kbit/s.

Connections from a host computer to the transmission medium is done using proprietary Allen-Bradley interface cards. The interfaces for DH485, the 1774 KR interface card and the 1747 isolated link coupler, are based on the RS485 standard for signaling. It employs one line for transmission, one line for reception, a signal and a chassis ground. Since the RS485 standard is designed for use on networks as opposed to point-to-point communication, there are no handshaking lines as there are in RS232. Transmission is controlled by the possession of the token.

Data Highway is capable of communicating in either a full duplex or half duplex mode. In full duplex mode, data highway can implement peer-to-peer communications between stations using a modified token passing scheme called a **floating master**. In this scheme, the stations on the network bid for possession of the token based on the need to send information. Half duplex is implemented using a **master/slave** scheme. In this arrangement, a computer serves as the supervisory controller and is the master on the network. It controls access to the token. Each processor (slave) on the network gains

COMMUNICATION ARCHITECTURE AND LANs

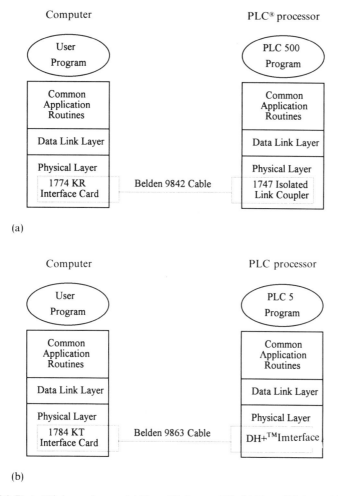

Fig. 7.16 Data Highway layers. (a) Data Highway 485; (b) Data Highway Plus™.

access to the token when the master sends the token to it. In the remainder of this section we will be discussing the communication protocol based on master/slave token passing scheme.

The data link layer employs a link protocol that ensures that the message is carried, intact, between nodes. The unit of transmission is the packet. When transmitting, it is responsibility of the data link layer to assemble the packet for transmission and to hand it off to the physical layer to transmit it. When receiving, it is the responsbility of the data link layer to accept the packet from the physical layer and check it for errors before sending it to the application layer.

There are three types of packets used in Data Highway communication using the master/slave token passing scheme: polling packet, master message

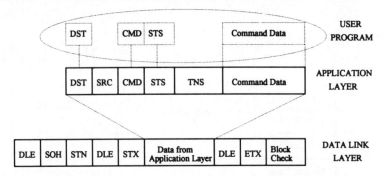

Fig. 7.17 Construction of Data Highway network master message packet.

packet and slave message packet. The polling packet is used to rotate the token around the network and provide fair access to each station. The host computer on the network is designated as the master. Transmission from the host is done using the master message packet; programmable controllers on the network transmit using the slave message packet. We shall examine the components of each of these packets.

A packet is broken down into two components: control symbols and data. Figure 7.17 shows the master message packet. The data link layer packet takes the data packet from the application layer and adds eight fields on to it. These fields and their purpose are as follows:

- DLE SOH: the first two fields are sender control symbols that indicate the start of a master message. By DH^{+TM} convention, DLE = $10 and SOH = $01.
- STN: This is the station number of the slave node to which the message is being sent. When a slave node receives DLE SOH, it prepares its buffer to latch the station number. If a match is found between STN and its assigned station number, it continues to process the message.
- DLE STX: these fields are control fields. They are used to separate the header information from the data field. By convention, STX = $02. When the slave node reads DLE STX, it prepares to latch the data string.
- DLE ETX: this field is a control field that indicates the termination of a message and separates the message packet from the control field. By convention, ETX = $03.
- Block check: this is the block check field used as one check of the validity of the transmission. The block check field is the 2s complement of the combination of the STN field and the data bits sent in the packet. Figure 7.18 illustrates the manner in which the block check is performed. Each frame of a hypothetical packet is shown with a hexadecimal representation of its content. The computation for the check is shown as the addition of the

COMMUNICATION ARCHITECTURE AND LANs 323

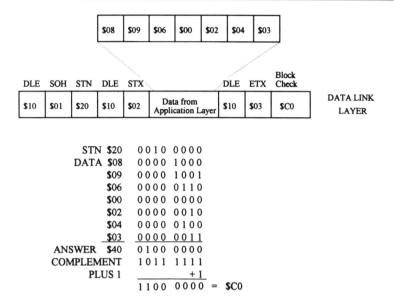

Fig. 7.18 Example of the use of the block check field.

station number ($20) and all the data field frames. The 2s complement of the result in $CO, which is transmitted in the Block Check field. When the receiving station evaluates the packet at the data link layer, it checks the block check field against the computation of Fig. 7.18. If they do not match, an error message is returned to the transmitting machine and a retransmission must take place.

Figure 7.17 shows the composition of the packet at the application layer and its relation to the data link layer. The reader will note that the added information at the data link layer are the control symbols, station number and block check field. In transmission, the data link layer adds these fields to the packet from the application layer. In reception, the data link layer ensures that the packet is for its address and performs the block check for data security. It then discards these fields and passes the remainder of the packet to the application layer.

The components of the packet at the application layer are as follows:

- DST: this field gives the destination node for the message.
- SRC: this field gives the source node, or transmitting station number. It is the station to which a reply packet will be sent when the received message is acknowledged or when a retransmission is requested.
- CMD: these fields define the activity to be done at the destination node. The command codes used in conjunction with the Allen-Bradley 5 series

Table 7.1 Partial Data Highway communication network command set

Command set	CMD code	FNC code
Basic command set:		
Diagnostic loop	$06	$00
Protected bit write	$02	N/A
Protected write	$00	N/A
Set timeout	$06	$04
Set variables	$06	$02
Unprotected read	$01	N/A
Unprotected write	$08	N/A
PLC 5 command set:		
Download all request	$0F	$50
Read bytes physical	$0F	$17
Upload all requests	$0F	$53
Word range read	$0F	$01
Word range write	$0F	$00
SLC 500 command set:		
Read link parameters	$06	$09
Set link parameters	$06	$0A
Protected logical read	$0F	$A2
Protected logical write	$0F	$AA

and 500 series programmable controllers are shown in Table 7.1 The CMD (command) frame defines the command type. The FNC (function) byte specifies the function under that command type. The FNC field is required in some commands, but not in others. When it is required, it is transmitted as part of the command data field. Two examples are shown in Fig. 7.19.

In Fig. 7.19a, an unprotected read is being requested. In an unprotected read, you can read the contents of the controller's memory where data is being stored. This memory is known as the data table. The address field specifies the starting address of the read and the size field specifies the number of sequential bytes to be read. The CMD byte 01 from the basic command set is sufficient to specify this operation. The reply packet format is also shown. When the unprotected read is accomplished, the processor returns the data in the reply format. The fact that it is a reply to a request for an unprotected read is indicated by reply CMD $41.

In Fig. 7.19b, a diagnostic loop is being executed. A diagnostic loop is used to check the integrity of the transmission link between stations. The master transmits up to 243 bytes in the data field. The receiving station replies with the same data field.

The other fields shown in Fig. 7.19 are interpreted as follows:

- STS: this is a status field. This field is always set to zero in the command mode. In the reply mode, a zero in this field indicates that the message has

COMMUNICATION ARCHITECTURE AND LANs

Command Format:

CMD $01	STS	TNS	ADDR	Size

Reply Format:

CMD $41	STS	TNS	DATA

(a)

Command Format:

CMD $06	STS	TNS	FNC $00	DATA

Reply Format:

CMD $46	STS	TNS	DATA

(b)

Fig. 7.19 Examples of application commands. (a) Unprotected read format; (b) diagnostic loop format.

been executed without error. A value in this field indicates an error has occurred.

- TNS: this is a two byte field called the transaction field. It is used to uniquely identify each command message. For each command message, the application layer software assigns a unique 16-bit transaction number which is usually assigned by indexing a 16-bit counter each time a message is sent. When the receiving node sends a reply, it copies the same transaction number into the TNS field of the reply. At any moment the combination of the SRC, CMD and TNS fields of a message is sufficient to uniquely identify every message packet.
- DATA: the data portion of the packet is specific to the command being sent. In Fig. 7.19 we have illustrated two cases. For the data fields associated with other commands, the reader is referred to the Allen-Bradley Data Highway Plus™/DH-485 Protocol reference manual.

Figure 7.20 shows the format of a slave message at both the data link and the application layers. Note the similarity to the master message link packet. The difference is the missing DLE SOH header, which is used only by the master to indicate that the message is from the master node.

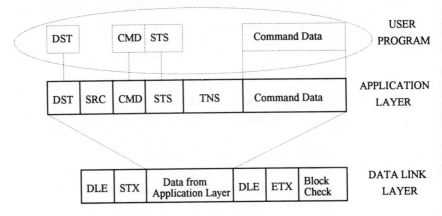

Fig. 7.20 Construction of slave message packet.

It is important to note that, from the point of view of the programmer, the entire process of assembling and transmitting packets is transparent. The physical, data link and application layers are a property of the communication network and are not directly manipulated by the programmer. As indicated in Fig. 7.16, the user program sits on top of this stack. The user program accesses the application layer by using C language functions provided by Allen-Bradley. Any of the functions of the command set can be called by a program written by the programmer. When a function call is made from the user software, the application layer begins to assemble the packet, which is passed to the data link layer for the addition of further fields, and finally to the physical layer for transmission. The user program is independent of the communication process except for calling the appropriate function.

Finally, Fig. 7.21 shows the format of the polling packet, which is a logical token. Recall that a node (master or slave) can send a message only when it is in possession of the token. Whenever a lull occurs in the transmission of messages, the poll packet is rotated around the network until a node on the network that is in possession of the token initiates a message. When the message is completed and acknowledged, the token is placed into rotation again. In this implementation, the poll packet is always passed from master to slave and back again. If the master does not receive the poll packet back from

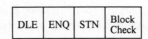

Fig. 7.21 Format of a polling packet.

the slave node within a specified period of time, it takes the slave node off the list of active nodes and continues to poll the other nodes. The poll packet has the following fields:

- DLE ENQ: this indicates that it is a transmission from the master and that it is a poll packet. ENQ is $05.
- STN: the station to which the token is sent.
- BCC: a block check, which is the 2s complement of STN.

This extended discussion of the Data Highway protocol leaves out much detail, for which the reader is referred to the Allen-Bradley manuals. Our interest here is to illustrate a fairly simple and widely used protocol in relation to the three levels of the OSI model commonly found in manufacturing LANs, in order to give some specific meaning to the way that the physical, data link and application layers function in an actual manufacturing implementation.

OSI was developed in anticipation of encouraging standards in networks. The primary example of an attempt at such standardization in manufacturing is the **Manufacturing Automation Protocol (MAP)** network standard. The MAP standard was encouraged by General Motors Corporation in the early 1980s. General Motors (GM) was a major purchaser of factory automation equipment. Working with its equipment vendors, they began to develop a standard that would allow machine tool controllers, robot controllers, PLCs and computers to communicate with each other over a common medium using a common protocol on the factory floor. Unfortunately, the MAP standard was not widely adopted outside of GM and the effort eventually dissipated. Most shop floor applications still use the proprietary networks of controller manufacturers and there is much point-to-point connectivity between machine controllers and supervisory computers using RS232. As was stated at the beginning of this section, the communication design problem is still one of the big difficulties in factory automation.

7.4 DISCRETE EVENT SYSTEMS AND SUPERVISORY CONTROLLER SOFTWARE DESIGN

In this section we focus on software design for the work cell and shop floor levels of control. In the metal fabrication industries and batch processing, these design problems are usually modeled as control problems for discrete event systems. A **discrete event system** is one in which the state of the system changes at discrete instances of time instead of continuously. Examples of this occur in many situation, e.g. the arrival of a part at a machine or the completion of the machining operation on a part. In each of these cases there is a change in the state of the system. In the first case the queue of parts at the machine goes from n to $n+1$; in the second case the completion of the machining operation allows the machine and operator to go from the state of machining to the state of unloading the part from the machine.

Although the PLC would appear to be the natural choice of controller for discrete events, this is not always the case. Depending on the amount of information that must be stored and processed and the control functions to be implemented, a more general purpose computer may be more appropriate. A conventional architecture for control hardware employs PLCs at the lower levels of control and shop floor computers at higher levels. An example of this was shown in Fig. 7.8. There we see a shop floor host computer that is responsible for all the supervisory activity on the factory floor. It is connected downstream to cell controllers over a LAN and is connected upstream to a factory host computer, on which resides the factory database that contains the permanently stored manufacturing information, such as schedules, available inventory, product routings and so forth. At the lower end of the network, the cell controllers, which may be PLCs or computers, are interfaced with various machines. They may be connected point-to-point over RS232 or the cell may be on a LAN. Regardless of the actual signaling standards that are used to communicate among subsystems, the general problem of analyzing the required supervisory control rules and developing the software control logic has to be addressed.

In the following sections we shall describe a general approach to discrete event system modeling and software design. We shall focus on the use of Petri nets as a tool for control software specification, analysis and implementation. Implementation of the control software specification can be done on a PLC or on a computer, as appropriate.

7.4.1 Petri net models and graphical representation

A Petri net (PN) is defined as a directed bipartite graph having a structure, C, and a marking, M. The structure, C, has four components: P, a set of places; T, a set of transitions; I, a set of input arcs that connect places to transitions; and O, a set of output arcs that connect transitions to places. Notationally, we use the definition $C = (\mathbf{P, T, I, O})$, where:

$\mathbf{P}$ = a set of places. $\mathbf{P} = \{p1, p2, p3, \ldots pn\}$
$\mathbf{T}$ = a set of transitions. $\mathbf{T} = \{t1, t2, t3 \ldots tm\}$
$\mathbf{I}$ = a mapping from places to transitions. $(\mathbf{PXT}) \to N$
$\mathbf{O}$ = a mapping from transitions to places. $(\mathbf{PXT}) \to N$

Each of these concepts has a graphical representation. The place is represented by a circle, the transition by a bar, and the input and output mapping by a set of directed arcs. Figure 7.22a is a Petri net. In this illustration there are four places, $\mathbf{P} = \{p1, p2, p3, p4\}$, and three transitions, $\mathbf{T} = \{t1, t2, t3\}$. There are also eight directed arcs which define $\mathbf{I}$ and $\mathbf{O}$. The input and output

DISCRETE EVENT SYSTEMS AND CONTROLLER SOFTWARE DESIGN

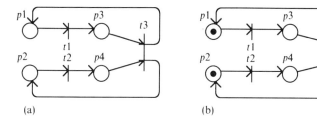

Fig. 7.22 Petri net, unmarked and marked.

mappings can be described in matrix form as follows:

$$\mathbf{I} = \begin{array}{c} \\ p1 \\ p2 \\ p3 \\ p4 \end{array} \begin{array}{ccc} t1 & t2 & t3 \\ 1 & 0 & 0 \\ 0 & 1 & 0 \\ 0 & 0 & 1 \\ 0 & 0 & 1 \end{array} \qquad \mathbf{O} = \begin{array}{c} \\ p1 \\ p2 \\ p3 \\ p4 \end{array} \begin{array}{ccc} t1 & t2 & t3 \\ 0 & 0 & 1 \\ 0 & 0 & 1 \\ 1 & 0 & 0 \\ 0 & 1 & 0 \end{array}$$

So, for example, t3 has two input places, p3 and p4. Therefore, the **I** matrix shows a 1 in the cells mapping those places to t3. Also t3 has two output places, p1 and p2. Therefore, the **O** matrix shows a 1 in the cells mapping those places to t3.

The four structural features, **P, T, I, O**, define the **topology** of the net. The **marking, M**, is a way of showing the current state of the net, where the state of the net defines which places are active. Formally, the marking vector is defined as $\mathbf{M} = \{m(p1), m(p1), m(p3), \ldots, m(pn)\}$. Graphically, it is symbolized by a token put in one or more places. When this is done, we refer to PN as a marked PN. Figure 7.22b is a marked PN in which p1 and p2 are active. Its marking vector is $\mathbf{M} = \{1, 1, 0, 0\}$.

Changes in the markings of a PN define its dynamics, or changes in state over time. The rules of Petri net dynamics are quite simple. A PN changes its state by changing its marking. A marking changes when a token moves from one place to another place. This is done by firing a transition. A transition may fire when it is enabled. A transition is enabled when there is a token in all of its input places. When the transition fires, it removes a token from each of its input places and adds a token to each of its output places. Figure 7.23 shows a firing sequence in which t1 fires first, followed by t2 and finally t3. In Fig. 7.23 the PN has passed through two states, returning to its initial state.

Note that the firing sequence shown in Fig. 7.23 is not the only possible firing sequence that could have occurred. For example, t2 could have fired before t1. We will return to this point in a later section.

The PN described above is called an **ordinary petri net**. In this class of PN, when a transition fires, one token is taken from each input place and one token

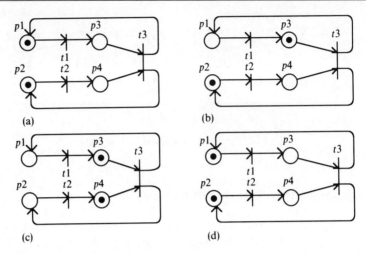

Fig. 7.23 Petri net dynamics. (a) Initial marking; (b) after firing $t1$; (c) after firing $t2$; (d) after firing $t3$.

is deposited into each output place. An ordinary Petri net has an arc weight of 1, where the arc weight indicates the number of tokens that traverse the arc when its transition fires. Figure 7.24 shows a PN with arc weights greater than 1. Here, when t1 fires, two tokens are removed from p1 and three tokens are added to p3. Also, the initial marking shows multiple tokens in a single place. The I and O matrices for Fig. 7.24, where the entries are the relevant arc weights are shown below. The entries of the marking vector represent the number of tokens in each place.

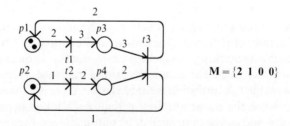

$M = \{2\ 1\ 0\ 0\}$

Fig. 7.24 A generalized Petri net with arc weights >1.

$$I = \begin{vmatrix} 2 & 0 & 0 \\ 0 & 1 & 0 \\ 0 & 0 & 3 \\ 0 & 0 & 2 \end{vmatrix} \quad Q = \begin{vmatrix} 0 & 0 & 2 \\ 0 & 0 & 1 \\ 3 & 0 & 0 \\ 0 & 2 & 0 \end{vmatrix}$$

DISCRETE EVENT SYSTEMS AND CONTROLLER SOFTWARE DESIGN

A PN can have any meaning, depending on the interpretation given it by the user. At this point it is convenient to think of the places as activities that occur over a period of time. The transitions can be thought of as events that occur at an instant in time.

By now the reader may have noted some similarities between a Petri net and the graphical programming language approach, which was described in Chapter 6. Both have transitions, and the transitions fire when their input states are enabled. In fact, GRAFCET was originally developed from Petri nets. It is a subclass of PN. In particular, it is a live, safe and repetitive Petri net. We shall discuss the meaning of some of these adjectives shortly.

Petri nets are more general than GRAFCET and are used to model more complex systems. This is shown in later examples. However, in order to illustrate fully the relationship to GRAFCET, we first apply PN to a familiar problem: the drill press example of section 6.9.1.

7.4.2 Petri net specification of a drill press work cycle

Recall from section 6.9.1 that the machining cycle consisted of two sub-cycles: the homing cycle and the work cycle. Figure 7.25 is a PN representation of the machine operations and is analogous to the GPL of Fig. 6.24. We can identify six states of the system. First, the machine enters the power on state at start-up. This state has the initial marking. From here the homing cycle transitions the machine to p3, where it waits for the first work cycle. Next, the drill head is descending (p4). Third, the drill head is ascending (p5). Finally, the drill head is idle waiting for the next cycle (p6). Figure 7.25 illustrates this sequence of states. Note the description of the transitions as events at an instant in time. This PN is but one interpretation of the activities of this machine. It should be noted that a simplification can be accomplished by eliminating p6 and having the firing of t6 place a token in p3.

Figure 7.25 describes a discrete event system. There are six discrete states during which we are not monitoring the operation of the system. Once the $-Z$ drive motor is turned on and the drill head is descending (state p4), we are not exercising control over the system until the drill head reaches the lower limit switch. At that point, the system must change its state (p4 → p5). In effect, we are only monitoring events that result in a change of state of the system. This contrasts with the case of continuous control, as in Chapter 5. If we were developing closed loop control software to regulate the speed of the $-Z$ drive motor, we would require continuous monitoring and regulation of the motor during state p4. This explains the difference between continuous and discrete event systems.

Figure 7.25 has given us a clear ordering of events for the drill press problem. With this model, we could study the behavior of the system as it evolves through each state. Figure 7.25 is known as an **autonomous Petri net** because it shows the evolution of the states without describing external events

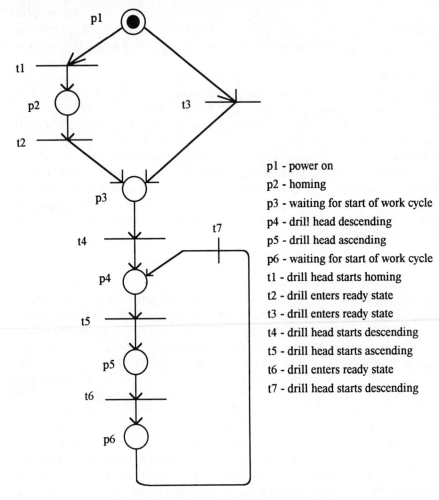

Fig. 7.25 Petri net model of drill press problem.

that condition the firing of transitions. In order to write the control logic, we must specify the events that lead to changes in the state of the system. When this is done, the result is a **non-autonomous Petri net**.

7.4.3 Adding external events

We shall follow a variation of the process used in Fig. 6.24 to annotate Fig. 7.25 and define external events. External events required for the firing of transitions will be labeled next to the transition and new outputs that become

DISCRETE EVENT SYSTEMS AND CONTROLLER SOFTWARE DESIGN

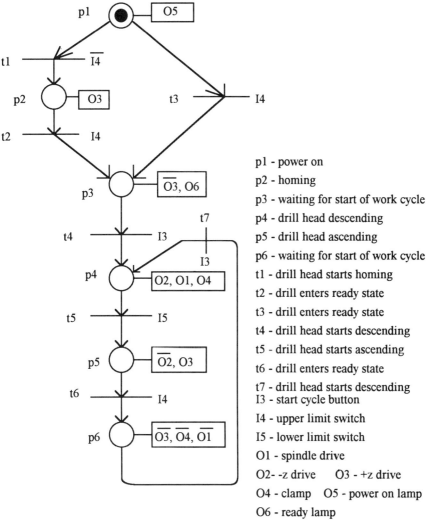

Fig. 7.26 Non-autonomous Petri net model of drill press problem.

activated when marking a place are labeled next to the place. The interpreted PN is shown in Fig. 7.26.

7.4.4 Software logic definition

Figure 7.26 has given us all the information required to write computer code. We begin by defining the logical relationships of the system. This is done by converting the annotated PN into a software logic definition by applying the same basic steps used in section 6.9.2, with a slight modification. This will involve writing three sets of equations and an initial marking:

1. transition firing equations
2. step activating and deactivating equations
3. output enabling and disabling equations
4. initial marking specification.

These four sets of specifications are as follows:

Transition firing equations:

$$p1 \cdot \overline{I4} = t1 \quad p3 \cdot I3 = t4$$
$$p2 \cdot I4 = t2 \quad p4 \cdot I5 = t5$$
$$p1 \cdot I4 = t3 \quad p5 \cdot I4 = t6 \quad p6 \cdot I3 = t7$$

Step activating and deactivating equations:

$$t1 = p1(U) \quad t3 = p1(U) \quad t5 = p4(U) \quad t7 = p6(U)$$
$$t1 = p2(L) \quad t3 = p3(L) \quad t5 = p5(L) \quad t7 = p4(L)$$
$$t2 = p2(U) \quad t4 = p3(U) \quad t6 = p5(U)$$
$$t2 = p3(L) \quad t4 = p4(L) \quad t6 = p6(L)$$

Output enabling/disabling equations:

$$p1 = O5(L) \quad p4 = O2(L) \quad p5 = O3(L)$$
$$p2 = O3(L) \quad p4 = O1(L) \quad p6 = O3(U)$$
$$p3 = O3(U) \quad p4 = O4(L) \quad p6 = O4(U)$$
$$p3 = O6(L) \quad p5 = O2(U) \quad p6 = O1(U)$$

Initial marking: $p1 = 1$; $pi = 0 (i = 2, 3, 4, 5, 6)$

If the reader compares the above equations with the ladder logic implementation of the graphical programming language model in section 6.9.2, it will be noted that the output enabling/disabling equation set has been changed. Here we have employed the use of the latch/unlatch functions. This was done in order to provide a more general set of equations, capable of dealing with more complex system behavior. In single machine control problems, the machine usually cycles through the same sequence of steps repetitively. For a simple problem, it is possibile to predict all the outputs that will be enabled during a step. In Fig. 6.24 we indicated all the outputs that would be enabled for each step. In more complex systems, there may be several paths through the network that will lead to the same state. Therefore, the outputs enabled during any particular state are a function of the pathway taken to arrive at that state. Hence, it is not always possible to decide which outputs will be activated when the token resides in a particular place.

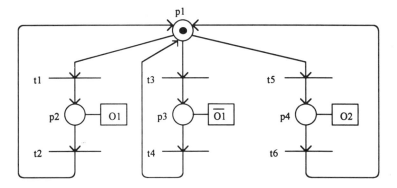

Fig. 7.27 PN structure with enabled outputs conditional on past events.

Figure 7.27 is an illustration of this point. The token in p1 can circulate in any combination of three circuits: t1→p2→t2; t3→p3→t4; t5→p4→t6. When p4 is marked, the state of O1 will depend on which of the place (p2 or p3) was previously marked. If p2 was last marked, O1 is enabled; if p3 was last marked, O1 is disabled. Therefore, it is not possible to incorporate the state of all outputs into any particular marking. This kind of generality in the execution of the system requires a more general set of equations, such as those given in the output enabling/disabling equations above.

7.4.5 Writing software code

The final step in the process is to create computer code from the above equations. This can be done in any language. In compiling control equations into code, we have previously employed the language of ladder logic. Following the procedures of section 6.9.2, it is left as an exercise for the reader to generate the control program.

7.5 SOME MATHEMATICAL PROPERTIES OF ORDINARY PETRI NETS

In section 7.4.2, the Petri net approach was used to define a particular problem structure so that it would be relatively easy to specify the input/outut relationships necessary to control the changes in state of the system. However, this is not the reason for which Petri nets were originally developed. It was originally developed as a basis for the modeling of communications between asynchronous components of a computer system. In such systems, parallelism or concurrency or events, asynchronous behaviour, and conflicts in accessing resources are quite common. PNs became a basis for modeling and analyzing the dynamics of such systems. Over the past couple of decades there have been many theoretical extensions to the modeling power of PNs. This includes the formulation of a mathematical framework in which different properties of a PN can be analyzed.

We should elaborate a bit on the concepts of concurrency (or parallelism) and conflict. Two transitions are said to be concurrent if they are causally independent, i.e. one transition may fire before or after or concurrently with the other. In Fig. 7.23, transitions t1 and t2 are in parallel. Note also that each place in Fig. 7.23 has only one input arc and only one output arc. This is a subclass of Petri nets known as a **marked graph**. A marked graph allows the modeling of concurrent events but does not allow the modeling of conflict.

Concurrent events are common in manufacturing systems. For example, two components may be moving through independent processes along two converyor lines. The state changes of each component should be modeled independently of each other. In Fig. 7.23, p1 and p2 can represent operations being performed on one component (token), while p2 and p4 are operations being performed on another component.

Conflict is illustrated in Fig. 7.28. Here the place p1 is input to two transitions. When p1 is marked with a token, both t1 and t2 are enabled; however, only one transition may fire. Hence, there is a conflict or choice whenever a place has more than one output arc. Note in Fig. 7.28 that each transition has only one input arc and only one output arc. This is subclass of Petri nets known as a **state machine**. A state machine can model conflict but cannot model concurrency.

Conflict is common in manufacturing systems. For example, if a component arriving into production can be machined on any of several machines, there exists a choice as to which machine shall get the component. In Fig. 7.28, the token in p1 can represent the component and p2 and p3 can represent machines capable of machining the component. The machine that gets the job is the one whose transition fires first.

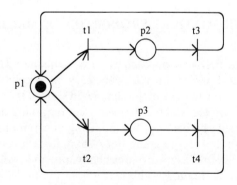

Fig. 7.28 PN with conflict.

SOME MATHEMATICAL PROPERTIES OF ORDINARY PETRI NETS

A PN can be constructed to model both concurrency and conflict at the same time. When it was recognized that concurrency, asynchronous behavior and resource conflict were characteristic of manufacturing systems, particularly discrete parts manufacturing systems, Petri nets were applied in modeling that arena. In this section we introduce some of this modeling power as a method for analyzing the behavior of a controller, which is a useful intermediate step between specifying the control system and actually generating code for use.

7.5.1 Mathematical modeling of Petri net dynamics

We have previously stated that the dynamic behavior of PN is defined by changes in its marking. The marking changes when a transition fires. A transition fires when its input states are marked. More formally, a transition tj is enabled in marking **M** if $M(pi) \geq I(pi, tj)$.

When a transition tj fires it results in a new marking, **M'**, which occurs by removing $I(pi, tj)$ tokens from each of its input places and adding $O(pi, tj)$ tokens to each of its output places. More formally, **M'** is reachable from **M** according to the equation:

$$M'(pi) = M(pi) + O(pi\ tj) - I(pi\ tj) \qquad (7.1)$$

So, for example, Fig. 7.29 shows a change in state from M_{k-1} to M_k by the firing of t1.

Let **u** be a firing vector, where $u^T = (u(t1), u(t2), u(t3))$. Then,

$$M_k = M_{k-1} + Ou_k - Iu_k$$

$$
\begin{array}{cccc}
M_k & M_{k-1} & O-I & u \\
\begin{vmatrix} 0 \\ 1 \\ 1 \\ 0 \end{vmatrix} = & \begin{vmatrix} 1 \\ 1 \\ 0 \\ 0 \end{vmatrix} + & \begin{vmatrix} -1 & 0 & 1 \\ 0 & -1 & 1 \\ 1 & 0 & -1 \\ 0 & 1 & -1 \end{vmatrix} & \begin{vmatrix} 1 \\ 0 \\ 0 \end{vmatrix}
\end{array}
$$

where a 1 in the firing vector indicates a firing of that transition. The matrix $O - I$ is an $n \times m$ matrix referred to as the incidence matrix, **A**. It defines the

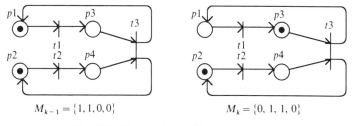

Fig. 7.29 Change in marking $M_{k-1} \rightarrow M_k$.

topology of the Petri net. The columns of **A** indicate the input places (-1) and output places (1) of each transition.

One can imagine a sequence of firings given by $u_1 + u_2 + u_3 + \ldots$ Therefore, to arrive at some destination marking, M_d, from an initial marking, M_o:

$$M_d = M_o + A \sum_{k=1}^{d} u_k$$

Let $\sum_{k=1}^{d} u_k = y$, $M_d - M_o = \Delta M$. Then

$$Ay = \Delta M, \tag{7.2}$$

and **y** is called the firing count vector. It is a vector whose elements are the number of times each transition fires in going from M_o to M_d. Note that the firing count vector only shows the number of times that each transition is fired in the sequence; it does not uniquely identify the sequence in which the firing takes place.

7.5.2 Petri net invariants

A number of important properties of a PN can be evaluated using the concept of PN invariants. An invariant of a PN depends on its topology. There are two invariants of interest, the P-invariant and the T-invariant.

Definition: If there exists a set of non-negative integers, **x**, such that $x^T A = 0$, then x is called a P-invariant of the PN.

Let **x** be a weighting vector of places, $x = [w_1 \; w_2 \; w_3 \ldots w_n]$. For the PN of Fig. 7.29 we make the computation

$$[w_1 \; w_2 \; w_3 \; w_4] \begin{vmatrix} -1 & 0 & 1 \\ 0 & -1 & 1 \\ 1 & 0 & -1 \\ 0 & 1 & -1 \end{vmatrix} = [0 \; 0 \; 0], \tag{7.3}$$

$$-w_1 + w_3 = 0$$
$$-w_2 + w_4 = 0$$
$$w_1 + w_2 - w_3 - w_4 = 0,$$

which has the following solutions:

$$\begin{array}{cccc} & p1 & p2 & p3 & p4 \\ x_1 = (& 1 & 0 & 1 & 0) \\ x_2 = (& 0 & 1 & 0 & 1) \\ x_3 = (& 1 & 1 & 1 & 1) \end{array}$$

Although the above three solutions exist, we are only interested in the *minimal* set of P-invariants. If an invariant is covered by two or more other

SOME MATHEMATICAL PROPERTIES OF ORDINARY PETRI NETS

invariants, it is not in the minimal set. Therefore x_1 and x_2 are minimal P-invariants and they cover x_3. In general there are $(n-r)$ minimal P-invariants, where n is the number of places and $r =$ the rank of $\mathbf{A}$.

The meaning of a P-invariant can be seen by observing Fig. 7.29. If one or more tokens exist in places p1 and/or p3, these two places will share the circulation of those tokens. This is disclosed by the invariant x_1. Similarly, places p2 and p4 may share the circulation of tokens. This is disclosed by the invariant x_2. The minimal P-invariants disclose the minimal structure of circuits that will share tokens. This also implies that these tokens are conserved within the circuit, i.e. the number of tokens in the set of places which make up the circuit, when weighted by the vector $\mathbf{x}$, is a constant.

Definition: A vector, $\mathbf{y}$, of non-negative integers is a T-invariant iff there exists a marking $\mathbf{M}$ and a firing sequence back to $\mathbf{M}$ whose firing count vector is $\mathbf{y}$.

Since $\mathbf{Ay} = \Delta \mathbf{M}$ as given in equation 7.2, a T-invariant is a solution to the equation $\mathbf{Ay} = 0$. Let $\mathbf{y}$ be a vector $\mathbf{y} = [u_1 \ u_2 \ldots u_m]$. For the PN of Fig. 7.29,

$$\begin{vmatrix} -1 & 0 & 1 \\ 0 & -1 & 1 \\ 1 & 0 & -1 \\ 0 & 1 & -1 \end{vmatrix} \begin{vmatrix} u_1 \\ u_2 \\ u_3 \end{vmatrix} = \begin{vmatrix} 0 \\ 0 \\ 0 \\ 0 \end{vmatrix}$$

$$-u_1 + u_3 = 0$$
$$-u_2 + u_3 = 0$$
$$u_1 - u_3 = 0$$
$$u_2 - u_3 = 0,$$

which yields the T-invariant $\mathbf{y}^T = (1\ 1\ 1)$. The meaning of the T-invariant is clear from equation 7.2. Since $\Delta \mathbf{M} = 0$, $\mathbf{y}$ defines the number of times each transition must fire in one complete cycle from $\mathbf{M}_0$ back to itself. In this instance, each transition will fire once during a cycle. In general, there are $(m-r)$ T-invariants, where m is the number of transitions.

The P and T-invariants of a PN provide tools to explore important properties of the system being modeled by the PN. In the following sections we examine some properties of a PN.

7.5.3 Reachability

It is often important to know what states a system can reach from some initial state. The set of all markings that can occur from some initial marking due to the firing of transitions defines the reachable states of the PN.

Property:

If a P-invariant $\mathbf{x}$ exists, then any reachable marking, $\mathbf{M}'$, from a marking $\mathbf{M}_o$, must satisfy the relation $\mathbf{x}^T\mathbf{M}' = \mathbf{x}^T\mathbf{M}_o$.

Proof:
$$\mathbf{M}' = \mathbf{M}_o + \mathbf{A}\mathbf{u}$$
$$\mathbf{x}^T\mathbf{M}' = \mathbf{x}^T\mathbf{M}_o + \mathbf{x}^T\mathbf{A}\mathbf{u}$$
since
$$\mathbf{x}^T\mathbf{A} = 0,$$
$$\mathbf{x}^T\mathbf{M}' = \mathbf{x}^T\mathbf{M}_o \tag{7.4}$$

EXAMPLE 7.1

With reference to Fig. 7.29, show that $\mathbf{M} = (0\ 0\ 1\ 1)$ is reachable from $\mathbf{M}_o$ but $\mathbf{M} = (1\ 0\ 1\ 0)$ is not reachable from $\mathbf{M}_o$.

Answer

Any reachable marking must satisfy equation 7.4, where $\mathbf{M}_o = (1\ 1\ 0\ 0)$.

Testing $\mathbf{M} = (0\ 0\ 1\ 1)$:

$$(1\ 0\ 1\ 0)\begin{vmatrix}0\\0\\1\\1\end{vmatrix} = (1\ 0\ 1\ 0)\begin{vmatrix}1\\1\\0\\0\end{vmatrix} \quad\quad (0\ 1\ 0\ 1)\begin{vmatrix}0\\0\\1\\1\end{vmatrix} = (0\ 1\ 0\ 1)\begin{vmatrix}1\\1\\0\\0\end{vmatrix}$$

$$1 = 1 \quad\quad\quad\quad\quad\quad 1 = 1$$

Testing $\mathbf{M} = (1\ 0\ 1\ 0)$:

$$(1\ 0\ 1\ 0)\begin{vmatrix}1\\0\\1\\0\end{vmatrix} \neq (1\ 0\ 1\ 0)\begin{vmatrix}1\\1\\0\\0\end{vmatrix}$$

$$2 \neq 1$$

The reachable markings can also be found by firing transitions until all the states of the system have been reached. This yields the **reachability graph** of the net, as shown in Fig. 7.30. This procedure for identifying reachable states is suitable for small problems, but can be unsatisfactory for large problems because the state space explodes.

SOME MATHEMATICAL PROPERTIES OF ORDINARY PETRI NETS | 341

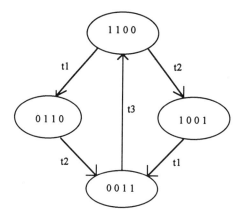

Fig. 7.30 Reachability graph for PN of Fig. 7.29.

7.5.4 Boundedness

A typical problem in manufacturing is to determine required buffer capacities between machining or assembly operations within the system in order to handle work in process. If tokens are used to represent parts being manufactured and places are used to represent buffers, it is of interest to know that the buffer capacities will not be exceeded. A PN is said to be k bounded if the number of tokens in each place does not exceed a finite number k for every reachable marking from $\mathbf{M}_o$. A PN is said to be safe if it is 1 bounded. For example, the PN of Fig. 7.28 is safe and the PN of Fig. 7.31 is unbounded.

In the above definition, boundedness depends on the initial marking. A stronger condition for boundedness is structural boundedness, which means that the PN is bounded for any finite marking $\mathbf{M}_o$.

Property:

A Petri net is structurally bounded iff there exists a non-zero vector, $\mathbf{x}$, of non-negative integers such that $\mathbf{x}^T\mathbf{A} \leq 0$.

Proof:

From equation (7.2) $\quad \mathbf{M} = \mathbf{M}_o + \mathbf{A}\mathbf{y}, \mathbf{y} \geq 0.$
Therefore, $\quad \mathbf{x}^T\mathbf{M} = \mathbf{x}^T\mathbf{M}_o + \mathbf{x}^T\mathbf{A}\mathbf{y}.$
Since $\mathbf{x}^T\mathbf{A} \leq 0$ and $\mathbf{y} \geq 0$, $\quad \mathbf{x}^T\mathbf{A}\mathbf{y} \leq 0$ and $\mathbf{x}^T\mathbf{M} \leq \mathbf{x}^T\mathbf{M}_o$
Therefore, for any marking, the number of tokens in all marked places has an upper bound as follows:

$$\sum_p \mathbf{x}_p^T \mathbf{M} \leq \mathbf{x}^T \mathbf{M}_o$$

Therefore, an upper bound for any individual place, p, is:

$$x(p)M(p) \leqslant x^T M_0$$

$$M(p) \leqslant (x^T M_0)[x(p)]^{-1},$$

where x(p) is the pth entry of **x**.

Another way to look at structural boundedness is to observe that a net becomes unbonded when there exists the condition that, when all its transitions fire, more tokens are added to output places than are absorbed from input places. Therefore, a necessary and sufficient condition for boundedness is that $\mathbf{x}^T\mathbf{O} \leqslant \mathbf{x}^T\mathbf{I}$, which yields the equation $\mathbf{x}^T\mathbf{A} \leqslant 0$.

Figure 7.31 is not structurally bounded. This is obvious since, for every firing of t1, tokens are sent to p2 as well as p3. Eventually, tokens will accumulate in the lower circuit without bound. Applying the criteria above,

$$\mathbf{x}^T\mathbf{A} \leqslant 0$$

$$|w_1 \; w_2 \; w_3 \; w_4| \begin{vmatrix} -1 & 0 & 1 \\ 1 & -1 & 1 \\ 1 & 0 & -1 \\ 0 & 1 & -1 \end{vmatrix} \leqslant |0 \; 0 \; 0|$$

which yields the inequalities:

$$-w_1 + w_2 + w_3 \leqslant 0$$
$$-w_2 + w_4 \leqslant 0$$
$$w_1 + w_2 - w_3 - w_4 \leqslant 0$$

The fact that there is no non-zero solution to the system of inequalities indicates the net is structurally unbounded for any initial marking. It should be pointed out that a structurally unbounded net may not be unbounded for a particular initial marking. Structural unboundedness refers to the fact that the topology of the net is such that there is at least one initial marking for which the net will be unbounded.

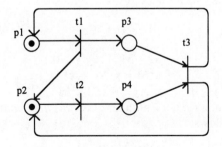

Fig. 7.31 Unbounded Petri net.

SOME MATHEMATICAL PROPERTIES OF ORDINARY PETRI NETS

In section 7.5.2 we discussed the fact that the P-invariant defined a circuit of the net in which tokens are conserved, i.e. the number of tokens in the places of the circuit are constant when weighted by the vector $\mathbf{x}$. By implication, if there exists a set of P-invariants that cover all places in PN, the weighted number of tokens in PN are constant. This would also imply that the net is bounded. Hence, another test of boundedness is to test for the conservative property, i.e. $\mathbf{x}^T \mathbf{A} = 0$, where x is a positive integer, $x \neq 0$.

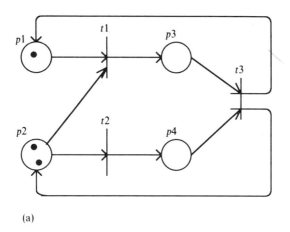

(a)

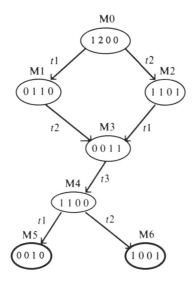

(b)

Fig. 7.32 PN with deadlock states.

7.5.5 Liveness and deadlock

An important property of any manufacturing system and its controller is that it will not enter a state of deadlock. System **deadlock** means that it has reached a state where it is not possible to continue to another state. A property associated with deadlock is **liveness**. A Petri net is live if, no matter what marking has been reached from M_o, it is possible to reach any transition in the net by progressing through some further firing. Obviously, if a PN is live, it is also deadlock free.

Unfortunately, there are no elegant mathematical methods for proving the existence of deadlock. Deadlock is found from the reachability graph of the net. If deadlock exists, the reachability graph will show a state (node) with no exist arc. Figure 7.30 is the reachability graph of Fig. 7.29. There are no deadlock states since each state has a transition arc to some other state. Figure 7.32a is a PN that contains a deadlock state. On the first cycle of events, t1 and t2 may fire because there are two tokens in p2. However, only one token is returned to p2. If t1 fires a second time, only p3 will be marked and the system is in deadlock. If t2 fires a second time, tokens will reside in p4 and p1 and the system will be in deadlock. Figure 7.32b is the reachability graph of the net and it clearly shows the deadlock conditions as states M5 and M6.

7.6 CASE STUDY: PETRI NET REPRESENTATION OF A MACHINING CELL CONTROLLER

In order to demonstrate some of the concepts of section 7.5, we introduce the problem of modeling a machining cell attended by a robot. The cell layout is shown in Fig. 7.33. The machining cell consists of a CNC lathe, a CNC milling machine and a robot for loading and unloading the machines. The lathe, milling machine and robot each have their own controllers to guide them through their machining cycle. The purpose of the cell controller is to sequence and coordinate the overall activities of the cell.

There are three buffer for holding parts. Parts are loaded from outside the cell into the input buffer. When a part is present and the lathe is available, the robot can load the lathe. When the lathe completes its machining cycle, the robot can unload the lathe and put the turned part into the WIP buffer. When the milling machine is available and a part is in the WIP buffer, the robot can load the mill. When the milling operation is complete, the robot unloads the machined part to the finished parts buffer where it is taken away by workers who clear the finished parts buffer when refilling the input buffer.

In constructing a model, we wish to model several aspects of the system. We wish to model the resources and their states. The resources are the robot, lathe and mill. We wish to model the flow of components as they move through the system. Finally, we wish to model the control logic for running the system.

PETRI NET REPRESENTATION OF A MACHINING CELL CONTROLLER

This cycle of events by which the system executes is modeled by the PN of Fig. 7.34. The cycle on the left characterizes various states of the lathe during the operating cycle of the cell and the cycle on the right characterizes various states of the mill during the operating cycle of the cell. The flow of material is also captured in the net. An input buffer either has a part available (IN) or it doesn't ($\overline{\text{IN}}$). The WIP inventory increases by 1 each time the lathe is unloaded (t5). The FP inventory is increased by 1 each time the mill is unloaded (t15). The PN also characterizes the availability of resources. The robot (R) is available when its place is occupied by a token. The lathe and mill could also be included in the net, but it is only necessary to show shared resources, such as the robot.

7.6.1 Testing boundedness

Following the discussion of section 7.5.4, we will test the condition for structural boundedness of the PN, namely that there exists a non-zero vector, $\mathbf{x}$, of non-negative integers such that $\mathbf{x}^T \mathbf{A} \leqslant 0$. The incidence matrix, $\mathbf{A}$, is written:

	t1	t2	t3	t4	t5	t6	t11	t12	t13	t14	t15	t16
p1	−1				1							
p2	1	−1										
p3		1	−1									
p4			1	−1								
p5				1	−1							
IN	−1					1						
$\overline{\text{IN}}$	1					−1						
WIP					1		−1					
p11							−1			1		
p12							1	−1				
p13								1	−1			
p14									1	−1		
p15										1	−1	
FP											1	−1
R	−1	1		−1	1		−1	1		−1	1	

Solving the set of linear inequalities $\mathbf{x}^T \mathbf{A} \leqslant 0$ may require an exhaustive search. A solution can be formulated as an integer program as follows:

$$\text{Min} \sum_i w_i$$

such that

$$\mathbf{x}^T \mathbf{A} \leqslant 0$$

$$w_i \geqslant 1 \text{ and integer.}$$

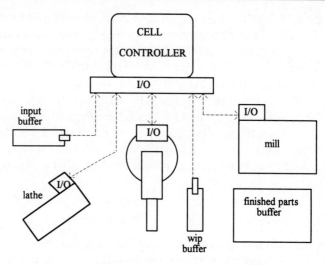

Fig. 7.33 Automated machining cell.

As suggested in section 7.5.4, another check on the boundedness condition can be made by checking whether or not the P-invariants cover the PN. This is done by solving for the minimal P-invariants using the relationship $\mathbf{x}^T \mathbf{A} = 0$. Let the weighting vector $\mathbf{x}$ be defined as $\mathbf{x} = [p1, p2, \ldots, FP, R]$. Then $\mathbf{x}^T \mathbf{A} = 0$ yields:

$$-p1 + p2 - IN + \overline{IN} - R = 0 \qquad (7.5.1)$$
$$-p2 + p3 + R = 0 \qquad (7.5.2)$$
$$-p3 + p4 = 0 \qquad (7.5.3)$$
$$-p4 + p5 - R = 0 \qquad (7.5.4)$$
$$+p1 - p5 + WIP + R = 0 \qquad (7.5.5)$$
$$+IN - \overline{IN} = 0 \qquad (7.5.6)$$
$$-WIP - p11 + p12 - R = 0 \qquad (7.5.7)$$
$$-p12 + p13 + R = 0 \qquad (7.5.8)$$
$$-p13 + p14 = 0 \qquad (7.5.9)$$
$$-p14 + p15 - R = 0 \qquad (7.5.10)$$
$$+p11 - p15 + FP + R = 0 \qquad (7.5.11)$$
$$-FP = 0 \qquad (7.5.12)$$

PETRI NET REPRESENTATION OF A MACHINING CELL CONTROLLER

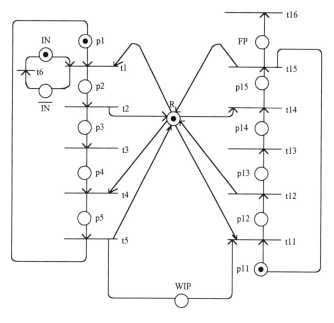

pl (p11)   Lathe (Mill) Available
p2 (p12)   Lathe (Mill) being loaded
p3 (p13)   Lathe (Mill) machining
p4 (p14)   Lathe (Mill) waiting to be unloaded
p5 (p15)   Lathe (Mill) being unloaded
IN ($\overline{\text{IN}}$)   Part available (not available)
WIP   Work in process buffer
FP   Finished parts buffer
R   Robot

t1 (t11)   Robot starts loading Lathe (Mill)
t2 (t12)   Robot completes loading Lathe (Mill)
t3 (t13)   Lathe (Mill) completes machining
t4 (t14)   Robot starts to unload Lathe
t5 (t15)   Robot completes unloading Lathe
t6 (t16)   Part enters (leaves) the system

Fig. 7.34 Petri net of automated machining cell of Fig. 7.33.

The above equations have the following solutions:

	pl	p2	p3	p4	p5	IN	$\overline{\text{IN}}$	WIP	p11	p12	p13	p14	p15	FP	R
$x_1 =$	(1	1	1	1	1	0	0	0	0	0	0	0	0	0	0)
$x_2 =$	(0	0	0	0	0	1	1	0	0	0	0	0	0	0	0)
$x_3 =$	(0	1	0	0	1	0	0	0	0	1	0	0	1	0	1)
$x_4 =$	(0	0	0	0	0	0	0	0	1	1	1	1	1	0	0)
$X =$	(1	2	1	1	2	1	1	0	1	2	1	1	2	0	1)

Recalling the interpretation of the P-invariant in section 7.5.2, places p1, p2,..., p5 share a token, as indicated by x_1. This is the token that circulates through the states of the lathe. Similarly, x_4 indicates the parallel activities of the mill. The P-invariant x_2 shows the two states of the input buffer; either a part is available or it is not available. Finally, x_3 shows the sharing of the robot. A token circulates among the state of robot availability and the loading and unloading of the lathe and the mill.

The sum of the P-invariants indicates that the PN is not covered. In particular, the WIP and FP inventories do not have a solution with positive non-zero integers. This can be quite easily seen for the case of FP by referring to equation 7.5.12, which has the solution $FP = 0$. An interpretation of this result is that WIP and FP may be unbounded. This can be seen by referring again to the PN of Fig. 7.34. Nothing prevents transition t5 from firing an infinite number of times before t11 fires. Thus, the WIP buffer can theoretically grow without bound. Hence, there is a fault in the logic of this model of the machining cell.

7.6.2 Modeling buffer capacities

In order to set bounds on the size of buffers, a simple structure for limiting capacity is used as shown in Fig. 7.35. In order for transition t5 to be enabled, a token must be present in the capacity place, $\overline{\text{WIP}}$. Therefore, t5 can fire only once before t11 must fire, limiting the WIP buffer capacity to one. When t11

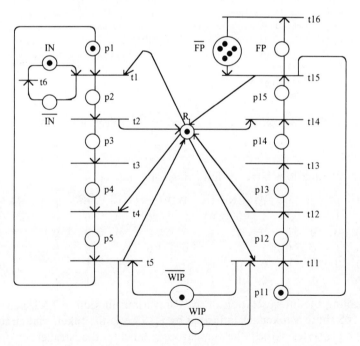

Fig. 7.35 Automated machining cell with buffer capacity contraints.

fires, a token is returned to the capacity place. A similar structure is implemented for the firing of t15. The capacity of FP is limited to five.

For Fig. 7.35, the solution to $\mathbf{x}^T \mathbf{A} = 0$ yields the following system of linear equations:

$$-p1 + p2 - IN + \overline{IN} - R = 0 \quad (7.6.1)$$
$$-p2 + p3 + R = 0 \quad (7.6.2)$$
$$-p3 + p4 = 0 \quad (7.6.3)$$
$$-p4 + p5 - R = 0 \quad (7.6.4)$$
$$-\overline{WIP} + p1 - p5 + WIP + R = 0 \quad (7.6.5)$$
$$+ IN - \overline{IN} = 0 \quad (7.6.6)$$
$$+\overline{WIP} - WIP - p11 + p12 - R = 0 \quad (7.6.7)$$
$$-p12 + p13 + R = 0 \quad (7.6.8)$$
$$-p13 + p14 = 0 \quad (7.6.9)$$
$$-p14 + p15 - R = 0 \quad (7.6.10)$$
$$-\overline{FP} + p11 - p15 + FP + R = 0 \quad (7.6.11)$$
$$+ \overline{FP} - FP = 0 \quad (7.6.12)$$

Equation set 7.6 differs from 7.5 only by the additions of $\overline{WIP}$ and $\overline{FP}$ in equations 7.6.5, 7.6.7, 7.6.11 and 7.6.12. Equation set 7.6 yields the following P-invariants:

	p1	p2	p3	p4	p5	IN	$\overline{IN}$	WIP	$\overline{WIP}$	p11	p12	p13	p14	p15	FP	$\overline{FP}$	R
$x_1 = ($	1	1	1	1	1	0	0	0	0	0	0	0	0	0	0	0	0)
$x_2 = ($	0	0	0	0	0	1	1	0	0	0	0	0	0	0	0	0	0)
$x_3 = ($	0	1	0	0	1	0	0	0	0	0	1	0	0	1	0	0	1)
$x_4 = ($	0	0	0	0	0	0	0	0	0	1	1	1	1	1	0	0	0)
$x_5 = ($	0	0	0	0	0	0	0	1	1	0	0	0	0	0	0	0	0)
$x_6 = ($	0	0	0	0	0	0	0	0	0	0	0	0	0	0	1	1	0)
$X = ($	1	2	1	1	2	1	1	1	1	1	2	1	1	2	1	1	1)

The addition of $\overline{WIP}$ and $\overline{FP}$ in the net has added two P-invariants, x_5 and x_6. The first indicates that the buffer and its complementary capacity place will share tokens; the second indicates the same situation for finished goods. The PN is now covered by P-invariants, indicating the net is bounded.

7.6.3 Testing for deadlock

To test for deadlock, we must enumerate the reachability graph from the initial marking of the PN. Even for the relatively small problem presented by this machining cell, the state space is rather large to be enumerated without

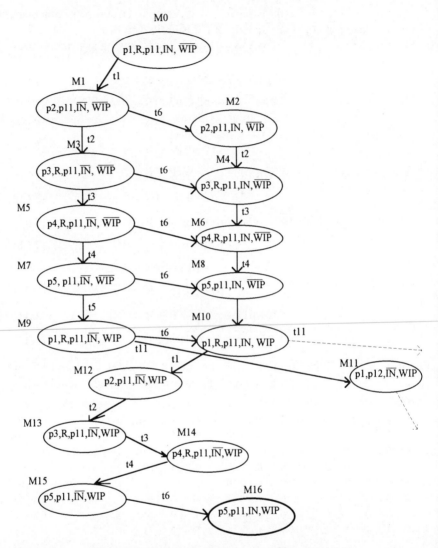

Fig. 7.36 Partial reachability graph of PN of Fig. 7.35.

computer assistance. In Fig. 7.36 a partial enumeration is given to the point at which a deadlock condition is found. To improve readability, each state node shows the places having a token at that point in the firing sequence. The deadlock condition exists in marking M16. At that marking the robot is in the process of unloading the lathe (p5), but there is no remaining space is the work-in-process buffer ($\overline{\mathrm{WIP}} = 0$). The reader may follow the events which have lead to M16 by working through the reachability graph. The main point is that the logic of the PN is still incomplete. There must be a means developed

PETRI NET REPRESENTATION OF A MACHINING CELL CONTROLLER

to prevent the system from entering state M16. This could be accomplished by preventing transition t4 from firing when there are no tokens in $\overline{\text{WIP}}$.

In Fig. 7.37, we illustrate two ways to prevent the deadlock state from occurring. In Fig. 7.37a the input arc of $\overline{\text{WIP}}$ has been moved from t5 to t4. This prevents the robot from entering p5 when there is no space in the buffer.

In Fig. 7.37b we show another modification of Fig. 7.35 by adding an arc from $\overline{\text{WIP}}$ to t4 and returning to itself. By so doing, we have conditioned t4 on the existence of a token in $\overline{\text{WIP}}$. Either of these modifications provides the logic necessary to prevent entering the deadlock state. It is equivalent to putting a sensor in WIP that indicates the buffer is full. The use of dashed lines in Fig. 7.37 has no special meaning; it is used to improve readability.

Although the logic of Fig. 7.37b is correct, there is a problem with analyzing the model mathematically. In particular, if $\overline{\text{WIP}}$ both enters and leaves t4, when we compute the incidence matrix, $\mathbf{A} = \mathbf{O} - \mathbf{I}$, the relationship we have just established will disappear. The same is true in the case of $\overline{\text{FP}}$. For the purpose of analysis, a PN must be pure. A PN is **pure** when there are no transitions having the same place as both an input place and an output place. To make Fig. 7.37b pure with respect to $\overline{\text{WIP}}$, it is necessary to add a dummy transition (td) and place (pd), as shown in Fig. 7.38. A similar modification should be made for $\overline{\text{FP}}$, which we will leave as an exercise for the reader.

Solving the relation $\mathbf{x}^T\mathbf{A} = 0$ for Fig. 7.38 yields the following equations:

$$-p1 + p2 - \text{IN} + \overline{\text{IN}} - R = 0 \qquad (7.7.1)$$

$$-p2 + p3 + R = 0 \qquad (7.7.2)$$

$$-p3 + p4 = 0 \qquad (7.7.3)$$

$$-p4 + p5 - R - \overline{\text{WIP}} + pd = 0 \qquad (7.7.4)$$

$$-\overline{\text{WIP}} + p1 - p5 + \text{WIP} + R = 0 \qquad (7.7.5)$$

$$+\text{IN} - \overline{\text{IN}} = 0 \qquad (7.7.6)$$

$$-pd + \overline{\text{WIP}} = 0 \qquad (7.7.7)$$

$$+\overline{\text{WIP}} - \text{WIP} - p11 + p12 - R = 0 \qquad (7.7.8)$$

$$-p12 + p13 + R = 0 \qquad (7.7.9)$$

$$-p13 + p14 = 0 \qquad (7.7.10)$$

$$-p14 + p15 - R = 0 \qquad (7.7.11)$$

$$-\overline{\text{FP}} + p11 - p15 + \text{FG} + R = 0 \qquad (7.7.12)$$

$$+\overline{\text{FP}} - \text{FP} = 0 \qquad (7.7.13)$$

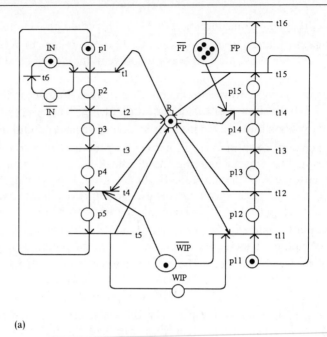

(a)

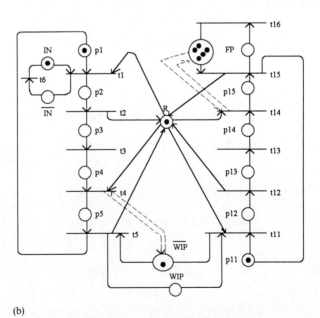

(b)

Fig. 7.37 Automated machining cell without deadlock.

SOFTWARE SPECIFICATION FOR MACHINING CELL CONTROLLER

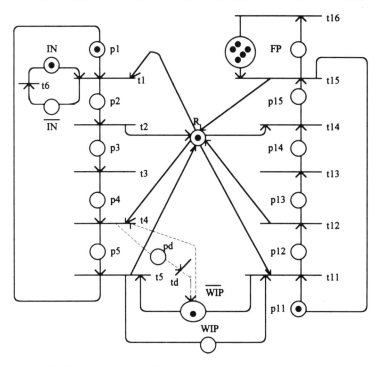

Fig. 7.38 Fig. 7.37 with additional transition and place to render pure.

The constraint on the firing of t4 is given in equation 7.7.4, which shows required inputs of p4, R, and $\overline{\text{WIP}}$. The new conditions introduced by equation 7.7.4 and 7.7.7 add one additional P-invariant in which pd and $\overline{\text{WIP}}$ will share tokens. At this point the model is free of the deadlock state at M16.

7.7 SOFTWARE SPECIFICATION FOR A MACHINING CELL CONTROLLER

In section 7.4.4 we introduced a method for developing the control program from an annotated Petri net by writing a set of equations that describes the evolution of the system. This is appropriate to the problem of specifying the drill press operation, but is not directly applicable to the machining cell problem as it is currently defined. A fundamental distinction exists between these two problems. In the drill press example, the controller was actually controlling the actuators of the machining cycle: spindle motor, clamp and z-axis drive motor. In the robot cell problem, the cell controller is not controlling actuators; rather, it is communicating with other controllers. The actuators of the lathe, the mill and the robot are being controlled by the controller of the lathe, mill and robot respectively. The cell controller is simply telling them when to begin their cycles. For the purposes of generating a control program for the cell controller, this fundamental difference must be taken into consideration by a proper specification of the communication protocol.

7.7.1 Petri net specification of communication protocols

Figure 7.39 is a simplified Petri net model of a communication protocol. In this figure there are two processes and one-way communication, from process 1 to process 2. When process 1 sends a message, it waits for an acknowledgement before it proceeds. When process 2 receives a message, it first acknowledges it before continuing on to execute further steps. In distributed control systems it is typical to include this handshake in the control logic in order to ensure that the intelligent device on the other end of the communication link has received the message. So, for example, when the lathe has been requested by the cell controller to begin machining, it is necessary for the cell controller to get an acknowledgement so that it can disable its request.

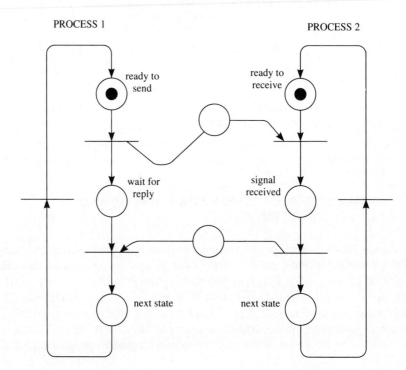

Fig. 7.39 PN model of a communication protocol.

SOFTWARE SPECIFICATION FOR MACHINING CELL CONTROLLER

7.7.2 Identification of communication requirements

In order to impose a communication protocol, it is first necessary to identify the signal requirements. There are four intelligent devices interacting: the cell controller, lathe controller, mill controller and robot controller. In addition, there are passive sensors providing input signals only and not requiring intelligent protocols. The signal requirements for each of these devices from the point of view of the cell controller are as shown in Table 7.2.

Figure 7.40 shows two proposed models of the interaction of the cell controller, shown in part on the right, with the lathe controller, shown in part on the left. In Fig. 7.40a, the controller starts in state p2 (lathe being loaded). When the cell controller wants the lathe to start machining, it asserts a request to 'start machining' (O1). The lathe controller then goes into the machining cycle and the cell controller marks state p3. In order for the cell controller to

Table 7.2 Signaling requirement for robot cell

Message	Cell controller output
Cell controller:	
Request to robot to load lathe	O0
Request to lathe to start machining	O1
Request to robot to unload lathe	O2
Request to robot to load mill	O3
Request to mill to start machining	O4
Request to robot to unload mill	O5
	Cell controller input
Robot controller:	
Robot available	I0
Robot loading lathe	I1
Robot unloading lathe	I2
Robot loading mill	I3
Robot unloading mill	I4
Lathe controller:	
Lathe available	I5
Lathe machining	I6
Mill controller:	
Mill available	I7
Mill machining	I8
Sensor signals:	
Input buffer, part available	I9
WIP, part available	I10
WIP, buffer full	I11
FP, buffer full	I12

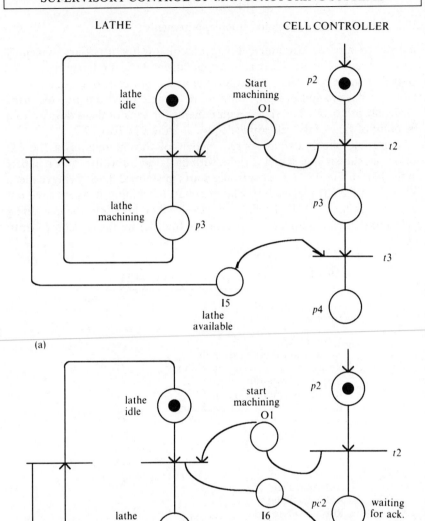

Fig. 7.40 Request for lathe machining cycle (a) without and (b) with handshake.

SOFTWARE SPECIFICATION FOR MACHINING CELL CONTROLLER

mark state p4 (lathe waiting to be unloaded), it must receive the signal I5 (lathe available). However, when that event occurs, the signal O1 is actually still on. That is to say, even though it would appear that enabling p3 in the lathe has absorbed the token in O1, nothing has told the cell controller to turn O1 off. Hence, as soon as the lathe is available again, the controller of the lathe immediately sees another request to begin machining. There is the possibility it will jump back into its machining cycle. This is an unsatisfactory arrangement and is precisely the reason for adding a handshake protocol when cell controller outputs represent communication signals as opposed to outputs that are directly operating dumb devices, such as clamps and motors. Figure 7.40b shows a proper protocol. As the lathe transitions into the machining state, it sends an acknowledgement signal back to the cell controller. At that point the cell controller knows it is time to disable O1. This requires the addition of a communication state, pc2, where the controller awaits the acknowledgement.

The general rule that applies is to add a communication handshake state to the cell controller each time it asserts a communication signal to an intelligent device. Note that this also has implications for all the intelligent devices (controllers) in the cell. The rules governing the protocol must be applied to all controllers in a consistent manner. In Fig. 7.40 we show only two states for the cell controller, which are the only two states relevant from the point of view of the lathe controller. However, the state 'lathe machining' is actually a number of states carried out by the lathe during its machining cycle. If we were to develop the control program for the lathe, these states would be made explicit.

7.7.3 Generating the control program

Figure 7.38 was used to analyze controller behavior and to identify information requirements for system operation. Table 7.2 includes these requirements. Figure 7.41 differs from Fig. 7.38 by replacing modeled buffer capacities with buffer capacity information.

Figure 7.41 is the interpreted PN with communication states. The rules of section 7.4.4 can be applied to define the software logic. Their application results in the following equations.

(a) Transition firing equations

$$p1 \cdot I0 \cdot I5 \cdot I9 = t1 \qquad p11 \cdot I0 \cdot I7 \cdot I10 = t11$$
$$pc1 \cdot I1 \cdot R = tc1 \qquad pc11 \cdot I3 \cdot R = tc11$$
$$p2 \cdot I0 = t2 \qquad p12 \cdot I0 = t12$$
$$pc2 \cdot I6 = tc2 \qquad pc12 \cdot I8 = tc12$$
$$p3 \cdot I5 = t3 \qquad p13 \cdot I7 = t13$$

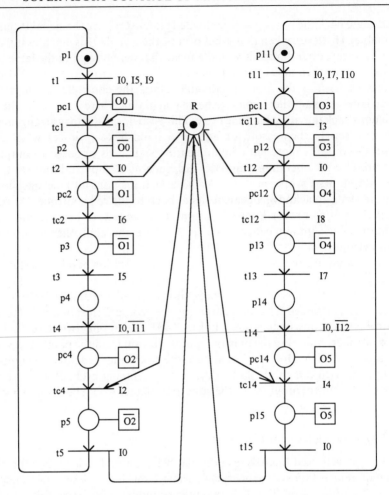

Fig. 7.41 Cell controller specification.

$$p4 \cdot I0 \cdot \overline{I11} = t4 \qquad p14 \cdot I0 \cdot \overline{I12} = t14$$
$$pc4 \cdot I2 \cdot R = tc4 \qquad pc14 \cdot I4 \cdot R = tc14$$
$$p5 \cdot I0 = t5 \qquad p15 \cdot I0 = t15$$

(b) Step activating/deactivating equations

t1 = p1(U)	t3 = p3(U)	t11 = p11(U)	t13 = p13(U)
t1 = pc1(L)	t3 = p4(L)	t11 = pc11(L)	t13 = p14(L)
tc1 = pc1(U)	t4 = p4(U)	tc11 = pc11(U)	t14 = p14(U)

tc1 = R(U)	t4 = pc4(L)	tc11 = R(U)	t14 = pc14(L)
tc1 = p2(L)	tc4 = pc4(U)	tc11 = p12(L)	tc14 = pc14(U)
t2 = p2(U)	tc4 = R(U)	t12 = p12(U)	tc14 = R(U)
t2 = pc2(L)	tc4 = p5(L)	t12 = pc12(L)	tc14 = p15(L)
t2 = R(L)	t5 = p1(L)	t12 = R(L)	t15 = p11(L)
tc2 = pc2(U)	t5 = R(L)	tc12 = pc12(U)	t15 = R(L)
tc2 = p3(L)	t5 = p5(U)	tc12 = p13(L)	t15 = p15(U)

(c) Output enabling/disabling equations

$$pc1 = O0(L) \quad pc4 = O2(L) \quad pc12 = O4(L)$$
$$p2 = O0(U) \quad p5 = O2(U) \quad p13 = O4(U)$$
$$pc2 = O1(L) \quad pc11 = O3(L) \quad pc14 = O5(L)$$
$$p3 = O1(U) \quad p12 = O3(U) \quad p15 = O5(U)$$

(d) Initial marking

$$p1 = p11 = R = 1;\ pi = 0,\ \text{otherwise.}$$

7.8 IMPOSING PRIORITIES IN PETRI NET MODELS

One of the important issues in implementing a controller is the avoidance or resolution of conflicts. There are two basic kinds of conflicts that occur in manufacturing systems: resource conflicts and task conflicts. **Resource conflicts** occur when more than one task requires a resource simultaneously. For example, if two jobs entering a manufacturing system require the same machine for processing, a resource conflict occurs. A **task conflict** occurs when more than one resource can be assigned to a given task, e.g. if one job entering a manufacturing system can be machined on any one of a number of workstations.

In developing a software specification, a common approach to establishing a priority is the use of an inhibitor arc. Figure 7.42 shows an inhibitor arc imposing a resource priority. In Fig. 7.42, two tasks, 'loading the lathe' and 'loading the mill', simultaneously require the robot resource. The inhibitor arc is drawn from the 'lathe available' place to the transition 'robot begins to load mill'. The round nob on the end of the inhibitor arc imposes a NOT condition on the transition. If a token exists in 'lathe available', the transition t11 cannot fire; if no token exists in 'lathe available', the transition is allowed to fire when other enabling conditions are present. The enabling equation for t11 is written:

$$t11 = p11 \cdot R \cdot \overline{p1}$$

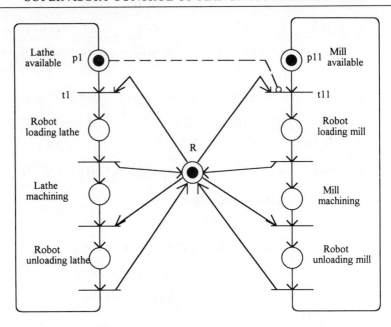

Fig. 7.42 Imposing a priority using an inhibitor arc.

Whenever the lathe is waiting to be loaded and the mill is waiting to be loaded at the same time, the robot will always load the lathe first.

The use of the inhibitor arc is sufficient to specify program logic. Once a PN model has been analyzed and is found to have acceptable behavioral properties, inhibitor arcs may be added before the controller code is generated. However, there is always some possibility that the addition of the inhibitor arc will change some behavioral property of the system that will not be detected unless the system is analyzed again with the inhibitor arc present. Unfortunately, the application of Petri net algebra to analyze behavioral properties does not include the use of inhibitor arcs.

The inhibitor arc priority structure can be modeled in another way that does allow the use of PN algebra. This requires the use of complementary places, an example of which is shown in Fig. 7.43. The complementary place, labelled cp, must have a token for t11 to fire. When t11 fires, a token is absorbed and immediately returned to cp. However, if t4 fires before t11 fires, the token in cp is absorbed into p1. This inhibits t11 from firing. When t1 fires, a token is returned to cp. This structure provides the same functionality as the inhibitor arc of Fig. 7.42. We will refer to this structure as a **priority net**.

IMPOSING PRIORITIES IN PETRI NET MODELS

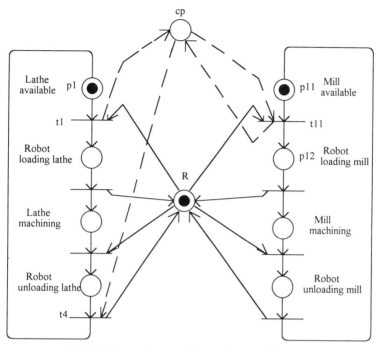

Fig. 7.43 Imposing a priority using a priority net.

7.8.1 Properties of priority nets

The basic properties of a priority net can be examined by looking at its basic structure, which is shown in Fig. 7.44. In Fig. 7.44a, a basic inhibitor arc structure is shown in which place pp (priority place) is imposing a priority on transition ti. Figure 7.44b shows the minimal structure for the corresponding priority net. We have added an additional place (p′) and transition (t′) in order to render the net pure, conserving its topological features, as previously discussed in section 7.6.3. Transition ti is the transition that is inhibited from firing if tp fires first. Place cp is the place that may inhibit the firing of ti. The priority net structure is defined by two places, cp and p′, a transition, t′, and their arcs.

We will assume that this minimal priority net structure is being imposed on some unpriorized net, let us call it the original net. Denote PN as the original net without a priority net and PN′ as the net after adding a priority net structure.

Property:

A priority net in PN′ will always add one additional P-invariant and no additional T-invariants to PN.

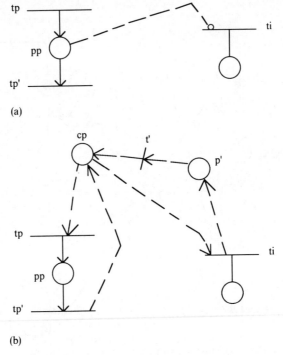

Fig. 7.44 Minimal structure for priority net.

Proof:

Consider the incidence matrix $\mathbf{A}^{n \times m}$ of PN and the incidence matrix $\mathbf{A}'$ of PN', as shown below. $\mathbf{A}$ is a submatrix of $\mathbf{A}'$, which is $(n+2) \times (m+1)$. If $\mathbf{A}$ is of rank r, $\mathbf{A}'$ is of rank $r+1$, since $cp = (pp + p')(-1)$ adds one additional independent equation. Since there are $n - r$ P-invariants and $m - r$ T-invariants in a PN, $\mathbf{A}'$ has $\Delta n - \Delta r = 2 - 1 = 1$ additional P-invariants and $\Delta m - \Delta r = 1 - 1 = 0$ additional T-invariants.

$$\mathbf{A}' = \begin{array}{c} \\ \\ \\ \\ \\ cp \\ p' \end{array} \begin{array}{|ccccc|c|} \hline & & & & & t' \\ a_{11} & \cdot & \cdot & \cdot & a_{1m} & 0 \\ \cdot & & & & \cdot & \cdot \\ \cdot & & \mathbf{A} & & \cdot & \cdot \\ \cdot & & & & \cdot & \cdot \\ a_{n1} & \cdot & \cdot & \cdot & a_{nm} & 0 \\ \hline \cdot & \cdot & \cdot & \cdot & \cdot & 1 \\ \cdot & \cdot & \cdot & \cdot & \cdot & -1 \\ \end{array}$$

IMPOSING PRIORITIES IN PETRI NET MODELS

$$\mathbf{A'} = \begin{array}{c} p_1 \\ \\ \\ \\ pp \\ \\ \\ \\ \\ \\ cp \\ p' \end{array} \begin{array}{cccccccccccccc} t1 & \cdot & \cdot & \cdot & ti & \cdot & \cdot & \cdot & tp & \cdot & \cdot & \cdot & tp' & \cdot & \cdot & \cdot & t' \\ a_{11} & & & & \cdot & & & & \cdot & & & & \cdot & & & & 0 \\ \cdot & & & & \cdot & & & & \cdot & & & & \cdot & & & & \cdot \\ \cdot & & & & \cdot & & & & \cdot & & & & \cdot & & & & \cdot \\ 0 & \cdot & \cdot & \cdot & 0 & \cdot & \cdot & \cdot & 1 & \cdot & \cdot & \cdot & -1 & \cdot & \cdot & \cdot & 0 \\ \cdot & & & & \cdot & & & & \cdot & & & & \cdot & & & & \cdot \\ \cdot & & & & \cdot & & & & \cdot & & & & \cdot & & & & \cdot \\ \cdot & & & & \cdot & & & & \cdot & & & & \cdot & & & & \cdot \\ \cdot & \cdot & \cdot & \cdot & -1 & \cdot & \cdot & \cdot & -1 & \cdot & \cdot & \cdot & 1 & \cdot & \cdot & \cdot & 1 \\ \cdot & \cdot & \cdot & \cdot & 1 & \cdot & \cdot & \cdot & 0 & \cdot & \cdot & \cdot & 0 & \cdot & \cdot & \cdot & -1 \end{array}$$

Property:

The prioritizing condition in PN' is given by the additional P-invariant.

Proof:

The additional equalities that $\mathbf{A'}$ adds to the invariants of $\mathbf{A}$ can be written as follows:

$$ti: cp = p' \quad t': p' = cp \quad tp: cp = pp \quad tp': pp = cp$$

From the above we see that $pp = 1 \Rightarrow cp = 1 \Rightarrow p' = 1$, which yields the minimal support P-invariants

$$\langle X_p \rangle = (pp \; cp \; p').$$

Since cp is a firing condition for ti and one token is being shared between pp, cp, and p', $pp = 1 \Rightarrow cp = 0$, and ti cannot fire.

Property:

For any priority net PN', $R'(M_o) \in R(M_o)$, where $R(M_o)$ is the reachability set, given an initial marking M_o, and R' denotes the markings of PN that are reachable in PN'.

Proof:

This follows logically from the first two properties. If $R(M_o)$ includes the state prohibited by the priority place in PN', then $R'(M_o)$ is a subset of $R(M_o)$. More formally, refer to the structure shown in Fig. 7.45, where p_{i-1} and p_{p-1} are input places of ti and tp respectively. The need for imposing a priority comes from the fact that pp = 1, p_i = 1, p_{i-1} = 0 is a reachable marking from pp = 1, p_{i-1} = 1, p_i = 0 in PN. If a marking vector, M_i, is reachable from M_{i-1}, it must satisfy the condition $M_i = M_{i-1} + Au$, where u is the firing vector. Applying this condition:

$$\Delta M' \qquad\qquad\qquad A' \qquad\qquad\qquad u'$$

	$\Delta M'$		t1	...	ti	...	tp	...	tp'	...	t'	
p1	Δm_1		a_{11}	··	·	··	·	··	·	··	·	u_1
·	·		·		·		·		·		·	·
·	·		·		·		·		·		·	·
·	·		·		·		·		·		·	u_{ti}
p_{i-1}	-1		0	··	-1	··	0	··	0	··	0	·
p_i	1	=	0	··	1	··	0	··	0	··	0	u_{tp}
·	·		·		·		·		·		·	·
p_{p-1}	0		0	··	0	··	-1	··	0	··	0	$u_{tp'}$
pp	0		0	··	0	··	1	··	-1	··	0	·
·	·		·		·		·		·		·	·
cp	0		0	··	-1	··	-1	··	1	··	1	$u_{t'}$
p'	1		0	··	1	··	0	··	0	··	-1	

The additional equalities in PN' can be written:

$$-1 = -u_{ti}$$
$$0 = -u_{tp}$$
$$1 = u_{ti}$$
$$0 = u_{tp} - u_{tp'}$$
$$0 = -u_{ti} - u_{tp} + u_{tp'} + u_{t'}$$
$$1 = u_{ti} - u_{t'}$$

Since there is no non-negative solution to the above equations, pp = 1, p_i = 1, p_{i-1} = 0 cannot be reached from pp = 1, p_{i-1} = 1, p_i = 0. Therefore, $R'(M_o) \in R(M_o)$. We note that the reachability set of PN', $R'(M_o)$, includes marked places not in PN due to the addition of places cp and p'.

IMPOSING PRIORITIES IN PETRI NET MODELS

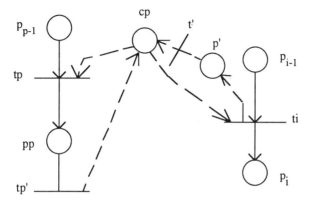

Fig. 7.45 Priority net and firing conditions.

7.8.2 Case study: Assembly cell with two robots

Figure 7.46 shows the structure of an example problem. Two robots are tending an assembly line. Two different subassembly operation activities (A1 and A2) take place at workstations WS1 and WS2. These subassemblies are queued at Q1 and Q2. When subassemblies can be mated, activity A3 produces a final assembly. Robot R2 performs activity A2 and robot R1 is responsible for A1 and A3.

Figure 7.47 shows the Petri net. Buffer sizes have been added to Q1 and Q2. The place M indicates the existence of an available mating of two subassemblies. Hence, M represents two subassemblies simultaneously moved from queue into workstation WS3. A priority rule is implemented in Figure 7.47: when tasks A1 and A3 are both competing for service, A3 is given priority. This means R1 will service A3 first.

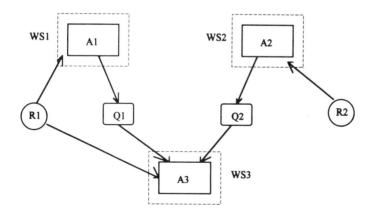

Fig. 7.46 Two robots tending assembly operations.

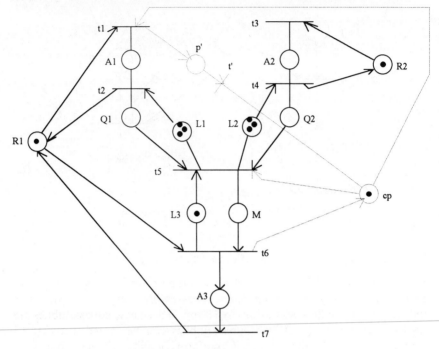

Fig. 7.47 Petri net for problem of Fig. 7.46.

The **A′** matrix with **A** submatrix is as follows:

	t1	t2	t3	t4	t5	t6	t7	t′
A1	1	−1	0	0	0	0	0	0
A2	0	0	1	−1	0	0	0	0
Q1	0	1	0	0	−1	0	0	0
L1	0	−1	0	0	1	0	0	0
Q2	0	0	0	1	−1	0	0	0
L2	0	0	0	−1	1	0	0	0
M	0	0	0	0	1	−1	0	0
L3	0	0	0	0	−1	1	0	0
A3	0	0	0	0	0	1	−1	0
R1	−1	1	0	0	0	−1	1	0
R2	0	0	−1	1	0	0	0	0
cp	−1	0	0	0	−1	1	0	1
p′	1	0	0	0	0	0	0	−1

It can be seen that matrix **A** is of rank 6 and **A**′ is of rank 7, and cp $= (-1)(p' + M)$. It can be shown that the P-invariants of **A** are:

A1	A2	Q1	L1	Q2	L2	M	L3	A3	R1	R2
(1	0	0	0	0	0	0	0	1	1	0)
(0	1	0	0	0	0	0	0	0	0	1)
(0	0	1	1	0	0	0	0	0	0	0)
(0	0	0	0	1	1	0	0	0	0	0)
(0	0	0	0	0	0	1	1	0	0	0)

This results in the following supports:

$$\langle X1 \rangle = \{A1 \ A3 \ R1\}$$
$$\langle X2 \rangle = \{A2 \ R2\}$$
$$\langle X3 \rangle = \{Q1 \ L1\}$$
$$\langle X4 \rangle = \{Q2 \ L2\}$$
$$\langle X5 \rangle = \{M \ L3\}$$

Similarly, the basic P-invariants and supports of **A**′ are:

A1	A2	Q1	L1	Q2	L2	M	L3	A3	R1	R2	cp	p′
(1	0	0	0	0	0	0	0	1	1	0	0	0)
(0	1	0	0	0	0	0	0	0	0	1	0	0)
(0	0	1	1	0	0	0	0	0	0	0	0	0)
(0	0	0	0	1	1	0	0	0	0	0	0	0)
(0	0	0	0	0	0	1	1	0	0	0	0	0)
(0	0	0	0	0	0	1	0	0	0	0	1	1)

$$\langle X1 \rangle = \{A1 \ A3 \ R1\}$$
$$\langle X2 \rangle = \{A2 \ R2\}$$
$$\langle X3 \rangle = \{Q1 \ L1\}$$
$$\langle X4 \rangle = \{Q2 \ L2\}$$
$$\langle X5 \rangle = \{M \ L3\}$$
$$\langle X6 \rangle = \{M \ cp \ p'\}$$

PN and PN′ are both structurally bounded since they are covered by invariants. Using the marking relationship $x^T M' = x^T M_o$ (equation 7.4), the first five supports yield:

$$m(A1) + m(A3) + m(R1) = 1 \quad (7.8.1)$$

$$m(A2) + m(R2) = 1 \quad (7.8.2)$$

$$m(Q1) + m(L1) = 3 \quad (7.8.3)$$

$$m(Q2) + m(L2) = 3 \quad (7.8.4)$$

$$m(M) + m(L3) = 1 \quad (7.8.5)$$

Equations 7.8.1 and 7.8.2 state that each robot can only work on one task at a time. Equations 7.8.3, 7.8.4 and 7.8.5 define the capacities of the subassembly locations. The final support provides the prioritizing condition:

$$m(M) + m(cp) + m(p') = 1, \quad (7.8.6)$$

$m(M) = 1 \Rightarrow m(cp) = 0$, and t1 cannot fire.

7.9 SUMMARY

Supervisory control refers to the control of a number of interacting unit operations in the manufacturing system. The supervisory control problem differs among different types of manufacturing system designs. In this chapter we have illustrated the problem for manufacturing cells in the discrete parts industry environment.

In computer automation of supervisory control, the factory local area network provides the means for communicating with unit operations and coordinating their activities. The message unit for communicating on a LAN is the message packet, which specifies message origin, message destination, and message content. In a minimal configuration, the LAN requires a physical layer to carry the packet, a data link layer to receive the packet and check its integrity, and an application layer to define its purpose. These layers were discussed in the context of a widely used manufacturing LAN, the Allen-Bradley Data Highway communication network.

A manufacturing system can be described as a discrete event system in which there is concurrency, conflict and asynchronous behavior. Such systems are very complex and present a challenge in modeling and developing control software. In this chapter we introduced the Petri net approach of specifying and modeling such systems. Methods were introduced for testing for important system characteristics, such as boundedness and deadlock. Setting priorities for conflict avoidance was illustrated using inhibitor arcs and priority nets. Finally, a procedure was suggested for translating a PN control specification directly into software logic for writing a control program.

In this chapter we have restricted our attention to ordinary Petri nets and their properties. There have been many extensions to Petri nets that enable the modeling of more complex problems. These include colored, stochastic and timed Petri nets. The bibliography below will direct the interested reader to the extensive literature on this subject.

EXERCISES

1. Consider the Petri net shown in Fig. 7.48.

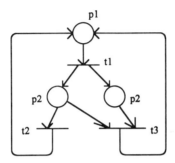

Fig. 7.48 PN for Exercise 1.

   (a) Solve for the P and T-invariants of the net.
   (b) Is the net structurally bounded?
   (c) Is the net deadlock free?

2. Consider the Petri net shown in Fig. 7.49.

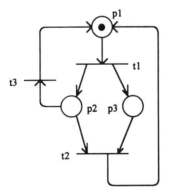

Fig. 7.49 PN for Exercise 2.

   (a) Develop the reachability graph of the net.
   (b) Compute the P and T-invariants of the net.

(c) Is the net structurally bounded?

(d) Are there any deadlock states?

3. Consider the matching cell of Fig. 7.6. Assume that all components are routed sequentially from the input buffer to lathe to mill to drill to output buffer. Components are not placed into work-in-process buffers. A part cannot be moved to the next machine until that machine is empty.

   (a) Develop a Petri net for this problem, being careful to prevent entering unwanted states. For example, a machine must release a part before a new part is loaded.
   (b) Define necessary input and output signals and write the equations of the system suitable for generating the control logic.

4. For Exercise 3 it is desirable to impose a rule that, when more than one machine requires service, the robot will always service the last machine in the sequence first. This is to avoid downstream machines from blocking throughput.

   (a) Define these priorities using inhibitor arcs.
   (b) Define necessary input and output signals and write the equations of the system suitable for generating the control logic.

5. For the situation of Exercise 4, replace the inhibitor arcs by priority nets.

   (a) Solve for the T and P-invariants of the system.
   (b) Show that the net is structurally bounded.

6. For the FMS of Fig. 7.7, assume that the FMS controller is dispatching the two AGVs to bring components to the three machining centers. Any machining center can machine any component. The buffer at each machining center has two locations (IN, OUT). The rules of dispatching are as follows:

   (i) When a machining center is not machining and its buffer is empty, an AVG is dispatched to bring a part to the machining center.
   (ii) If the output buffer is occupied, an AVG is dispatched to bring a component to the input buffer of the machining center and, at the same time, retrieve the finished component from the output buffer.

   Prepare a Petri net model of the system operation. Assume that each machining cell has its own controller that operates its respective cell once a part appears in its input buffer.

7. For the situation of Exercise 6, assume that one of the cells of the FMS is the robot cell described in Fig. 7.38.

   (a) Show how the Petri net of Exercise 6 is related to Fig. 7.38 if both nets were to be shown on the same graph.
   (b) From a control logic perspective, does your answer to part (a) indicate

that the FMS controller and cell controller should be operating independently or in conjunction with each other? Explain.
(c) Would your answer to part (b) change if the FMS controller were given the responsibility of initiating the action of the machining center (by downloading a start command)?

FURTHER READING

Allen-Bradley (1991) *Data Highway/Data Highway Plus^{TM}/DH-485 Communication Protocol and Command Set Reference Manual*, Allen-Bradley Company, Inc., Milwaukee, Wisconsin.

Black, J. T. (1991) *The Design of the Factory with a Future*, McGraw-Hill, Inc., New York.

Boucher, T. O., Jafari, M. A. and Meredith, G. A. (1990) Petri net control of an automated manufacturing cell, *Advanced Manufacturing Engineering*, Vol. 2, No. 4.

Boucher, T. O. and M. A. Jafari (1992) Design of a factory floor sequence controller from a high level system specification, *Journal of Manufacturing Systems*, Vol. 11 No. 6.

David, R. and Alla, H. (1992) *Petri Nets and Grafcet*, Prentice Hall International, Hertfordshire, United Kingdom.

Desrochers, A. A. and Al-Jaar, R. Y. (1995) *Applications of Petri Nets in Manufacturing Systems: Modeling, Control, and Performance Analysis*, IEEE Press, Piscataway, New Jersey.

DiCesare, F., Harhalakis, G., Proth, J. M., Silva, M. and Vernadat, F. (1993) *Practice of Petri Nets in Manufacturing*, Chapman & Hall, London.

Jafari, M. A. and Boucher, T. O. (1994) A rule based system for generating a ladder logic control program from a high level systems models, *Journal of Intelligent Manufacturing*, Vol. 5, No. 2.

Jafari, M. A., Meredith, G. A. and Boucher, T. O. (1995) A transformation from a boolean equation control specification to a Petri net, *IIE Transactions*, Vol. 27, No. 1.

Murata, T. (1989) Petri nets: Properties, analysis and applications, *Proceedings of IEEE*, Vol. 77, No. 4.

Rodd, M. G. and Deravi, F. (1989) *Communication Systems for Industrial Automation*, Prentice Hall International, Hertfordshire, UK.

Schatt, S. (1987) *Understanding Local Area Networks*, Howard W. Sams & Co., Indianapolis, Indiana.

Stallings, W. (1990), *Local Networks*, 3rd edn, Macmilian Publishing Company, New York.

Appendix: Conversion tables

Table A.1 Common weights and measures

Imperial	Metric	Metric	Imperial
Length			
1 inch (in)	25.4 millimetres (mm)	1 millimetre	0.039 inch
1 foot (ft)	0.3048 metre (m)	1 centimetre (cm)	0.394 inch
1 yard (yd)	0.9144 metre (m)	1 metre (m)	1.094 yards
1 mile	1.609 kilometres (km)	1 kilometre	0.6214 mile
Area			
1 square inch (in^2)	6.45 square centimetres (cm^2)	1 square centimetre	0.155 in^2
1 square foot (ft^2)	0.929 square metres (m^2)	1 square metre	10.764 ft^2
Volume			
1 fluid ounce (fl oz)	28.413 cubic centimetres (cm^3)	1 cubic centimetre	0.0352 fl oz
		1 cubic centimetre	0.061 in^3
1 gallon (gal)	4.546 litres (l)	1 litre	0.2199 gallon
1 cubic inch (in^3)	16.4 cubic centimetres		
Weight			
1 ounce (oz)	28.35 grams (g)	1 grams	0.035 oz
1 pound (lb)	0.454 kilogram (kg)	1 kilogram	2.205 lb
Velocity and acceleration			
1 ft/s	30.48 cm/s	1 cm/s	0.0328 ft/s
1 ft/s^2	0.3048 m/s^2	1 m/s^2	3.281 ft/s^2
Force			
1 lb	4.45 newtons	1 newton	0.2247 lb
Moment of inertia			
1 oz-in-s^2	7.06155×10^{-3} kg m^2	1 kg m^2	141.612 oz-in-s^2
Torque			
1 oz-in	7.06155×10^{-3} Nm	1 Nm	141.612 oz-in
1 lb-ft	1.35582 Nm	1 Nm	0.737562 lb-ft

Index

ASCII code 69
Actuator
 DC motor 148–58, 174–97
 flapper valve 167
 pneumatic cylinder 164
 relay switch 162
 solenoid 164
 solenoid valve 166
 stepping motor 158–62
Aikens, Howard 4
Analog to Digital (A/D) converter 100, 106, 111–16
Assembly cell 365
Assembly language programming 57–60
 Intel 8080 64
 Motorola 6800 58, 65
Automatic factory 6
Automation 7

Babbage, Charles 4
Boolean algebra 27–9, 257

Clock cycle 39
Communications 60–80
 band rate 69
 continuous polling 75
 interrupt driven systems 77
 parallel ports 65–9
 periodic polling 75
 RS232 71–4
 serial ports 69–74
 see also Local area network

Computer architecture 31–41
 accumulator 37, 59
 arithmetic logic unit 37
 address bus 33
 control memory 37
 data bus 33
 decoder 35
 instruction register 37
 internal stack 38
 memory address register 37
 memory data register 37
 program counter 37
 stock pointer 38
Computer memory 32–4
 Erasable programmable read only memory (EPROM) 33
 Random access memory (RAM) 32
 Read only memory (ROM) 32–4
Control Theory
 block diagram algebra 203
 characteristic equation 199
 closed loop 9, 186
 continuous control 10, 172
 control strategy 205 8
 critically damped 200
 digital control 213–29
 discrete control 10
 first order process 181
 open loop 9
 overdamped 200
 Proportional-integral-derivative (PID) 206
 second order process 200
 simulation 214–18
 steady state error 187
 steady state period analysis 202
 summary point 172
 transient period analysis 198–202
 underdamped 200

Damping 144, 182
Digital to Analog (D/A) converter 107–11
Discrete event system 327

Fail safe condition 255
Final value theorem 203
Flip flops 42
 D flip flop 44–7
 JK flip flop 42–4
Ford, Henry 3

INDEX

GRAFCET 271–86
Gear ratio 145

Inertia 141
  of lead screw 147
  reflected 146
Integrated circuit 16

Ladder logic 251
  arithmetic instructions 288
  bit instruction 251
  clear instruction 289
  conditional actions 289
  counters 263–6
  input instruction 252
  logical operations 288
  move instruction 288
  output instruction 253
  timers 267–71
Laplace transforms 176–9
  table of 177, 233
Lead screw 147
Local area networks 308–27
  Allen-Bradley Data Highway
    communication network 320–27
  bridges 319
  ethernet 313
  gateways 318
  manufacturing automation
    protocol 327
  network topology 310
  OSI reference model 314–20
  routers 316
  transmission medium 312
Logic gate 23
  AND gate 23
  EOR gate 30
  NAND gate 25
  NOR gate 26
  NOT gate 25
  OR gate 24
Logic networks
  combinatorial 29
  sequential 42

Machine cycle 38–41
Machining cell 306, 344–59
Manufacturing systems 301–8
  batch process 302
  continuous process 301
  flow line 304
  FMS 306

group technology 304
job shop 303
Mechanization 2
Melman, Seymour 4
Memory map 64, 287
Microcomputer 13
Microcontroller 15
  Motorola 68HC11 15, 64
Microprocessor 13
  Motorola 6800 59, 78
  Intel 8080 64
Move profile 144

Number systems 47–57
  2's complement 52
  binary 48
  binary coded decimal 54
  hexadecimal 55

Partial fraction expansion 230
Petri nets 328–68
  boundedness property 341–5
  communication protocols 354
  deadlock 344, 349
  marked graph 336
  marking vector 329
  ordinary, Petri net 329
  P-invariant 338
  priority net 359–68
  reachability property 340
  state machine 336
  T-invariant 339
  topology 329
Programmable logic controller (PLC)
  15, 240, 251–90
  AB PLC 5 321
  AB series 500 15, 321
  continuous control 290
  image tables 258
  memory map 287
  scan sequence 258
Programmed data transfer 62
Pulse width modulation 157

Registers
  memory 45
  serial to parallel 46
Relay ladder diagrams 244–51

Sampling process 218–21
Sensors
  float transducer 122

INDEX

flow rate   125
fluid flow switch   100
linear variable differential
  transformer (LVDT)   117
load cell   127
mechanical limit switch   93
optical encoder   120, 124
photoelectric   97
potentiometer   103, 117
proximity switch   95
resistance–temperature detector
  (RTD)   131
resolver   119
sensor array   99
strain gage   128
tachometer   124
thermister   131
thermocouple   130
ultrasonic range sensor   123

Sequential function chart (SFC)   271–86
Shannon, Claude   5
Supervisory control   8, 301

Time constant   181, 200
Torque   139–44, 174
  of lead screw   147
  reflected   145
Torque/speed curve   153, 156, 181
Transfer function   174
Transistor   16–23

von Neumann, John   5

z transform   224
  table of   225
Zero order hold   221–4